Student Solutions Manual

Calculus
with
Applications

Student Solutions Manual

Louis A. Guillou
Saint Mary's College of Minnesota

Calculus
with
Applications

Dale Varberg
Hamline University

Walter Fleming
Hamline University

Prentice Hall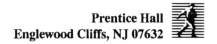
Englewood Cliffs, NJ 07632

Manufacturing buyers: Paula Massenaro and Lori Bulwin
Supplement Acquisitions Editor: Alison Munoz
Acquisitions Editor: Steve Conmy

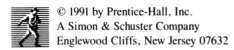
Printed in the United States of America

10 9 8 7 6 5 4 3 2 1

ISBN   0-13-115718-3

Prentice-Hall International (UK) Limited, *London*
Prentice-Hall of Australia Pty. Limited, *Sydney*
Prentice-Hall Canada Inc., *Toronto*
Prentice-Hall Hispanoamericana, S.A., *Mexico*
Prentice-Hall of India Private Limited, *New Delhi*
Prentice-Hall of Japan, Inc., *Tokyo*
Simon & Schuster Asia Pte. Ltd., *Singapore*
Editora Prentice-Hall do Brasil, Ltda., *Rio de Janeiro*

CONTENTS

NOTE TO THE STUDENT

This manual contains solutions to problems whose numbers are divisible by three. It was prepared for the purpose of helping you to carefully write solutions that will in turn help you to learn and review important ideas, definitions, theorems, and applications of calculus.

In the early chapters most steps are provided. As you progress and your mathematical skills develop you will be expected to fill in more steps. By the end of the manual, only major steps are provided.

Make it your goal to become independent of this manual. Only use it after you have tried a problem and are "stuck," and then only use enough of the solution to get you started. It is important that you learn to "do," not just be able to see that someone else's solution is correct.

Finally, a request. If you find any errors, typographically or mathematical, or if you have suggestions for improved solutions, please contact me at the Department of Mathematics & Statistics, St. Mary's College of Minnesota, Winona, MN 55987. Thank you.

L.G.

ACKNOWLEDGEMENTS

The problems were also solved by the authors. Their work was an excellent aid for checking solutions as well as for providing improved solutions for some problems.

The graphs in this manual were produced by my son, Tom. Additional invaluable assistance and support were provided at various stages by my wife, Maureen, by my sons, Mike, Joe, and Dan, and by Leah Gneiting. Their support is gratefully acknowledged.

Finally, I wish to thank the Prentice-Hall staff for their patience and support throughout the project.

Student Solutions Manual

Calculus
with
Applications

Problem Set 1.1   Functions

3. $V = (x)(x)(2x)$
    $V = 2x^3$

6. (a) $g(1) = 5(1) + 2 = 7$
    (b) $g(-1) = 5(-1) + 2 = -3$
    (c) $g(10) = 5(10) + 2 = 52$
    (d) $g(-0.4) = 5(-0.4) + 2 = 0$
    (e) $g(0.3) = 5(0.3) + 2 = 3.5$

9. (a) $f(2) = 3(2) - 2(2)^2 = -2$

    (b) $f(2+h) = 3(2+h) - 2(2+h)^2$
    $= -2h^2 - 5h - 2$

    (c) $\dfrac{f(2+h) - f(2)}{h} = \dfrac{(-2h^2 - 5h - 2) - (-2)}{h} = \dfrac{-2h^2 - 5h}{h} = -2h - 5 \quad (h \neq 0)$

    (d) $\dfrac{f(x+d) - f(x)}{d} = \dfrac{[3(x+d) - 2(x+d)^2] - [3x - 2x^2]}{d} = 3 - 4x - 2d \quad (d \neq 0)$

12. $\dfrac{f(a+h) - f(a)}{h} = \dfrac{\dfrac{2}{(a+h)^2} - \dfrac{2}{a^2}}{h} = \dfrac{2a^2 - 2(a+h)^2}{a^2(a+h)^2 h} = \dfrac{-4a - 2h}{a^2(a+h)^2} \quad (h \neq 0)$

15. (a) $g(f(1)) = g(-2) = 0.5$
    (b) $g(f(2.5)) = g(1) = 2$
    (c) $g(f(2)) = g(0)$ is undefined
    (d) $g(f(x)) = g(2x-4) = 2(2x-4)^{-2}$
    (e) $g(g(x)) = g(2x^{-2}) = 2[(2x^{-2})^{-2}] = \frac{1}{2} x^4 \quad (x \neq 0)$

18. (a) $g(f(1)) = g(2) = 4$
    (b) $g(f(4)) = g(2.5) = 6.25$

    (c) $g(f(2)) = g(\sqrt{2} + 1/\sqrt{2}) = 4.5$
    (d) $g(f(x)) = g(\sqrt{x} + 1/\sqrt{x})$

    $= (\sqrt{x} + 1/\sqrt{x})^2 = x^{-1}(x+1)^2$

21. $g(f(x)) = g(x^3+1) = (x^3+1)^2 = x^6 + 2x^3 + 1; \ f(g(x)) = f(x^2) = (x^2)^3 + 1 = x^6 + 1$

24. $h(g(f(x))) = h(g(x^{-2})) = h(4(x^{-2})^2) = h(4x^{-4}) = \sqrt{4x^{-4}} = 2x^{-2}$

    $f(g(h(x))) = f(g(\sqrt{x})) = f(4(\sqrt{x})^2) = f(4x) = (4x)^{-2} = \frac{1}{16} x^{-2}$

27. $f(x) = \ell nx; \ g(x) = cosx; \ h(x) = \sqrt{x}$

30. (a) $f(-3) = 4$ (b) $g(25/4) = 5/2$
     (c) $h(5/3) = 9/34$ (d) $h(g(f(0))) = h(g(1)) = h(1) = 1/2$
     (e) $f(g(h(\sqrt{5}/2) - f(g(4/9)) = f(2/3) = 25/9$
     (f) $h(g(f(x))) = h(g((x+1)^2)) = h(|x+1|) = (x^2+2x+2)^{-1}$
     (g) $f(f(x)) = f((x+1)^2) = [(x+1)^2 + 1] = (x^2+2x+2) = x^4+4x^3+8x^2+8x+4$

33. $f(x) = 700+60x, \ x \geq 0$

36. $V(x) = (14-2x)^2(x) = 4x(x-7)^2$.
The domain is the interval $(0,7)$.

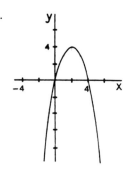

39. (a) $p(0) = 0; \ p(3) = 119700/2197 \approx 54.48$
     $p(20) = 2600/27 \approx 96.30$

     (b) $99 = 100\left[1 - \left(\dfrac{10}{10+t}\right)^3\right]$; $t = 10(0.01)^{-1/3} \approx 36.4$, so 99% will be discharged within about 37 days.

Problem Set 1.2 Graphs of Functions

3. (a) Odd; $f(-x) = 5(-x)^3-7(-x) = -5x^3+7x = -(5x^3-7x) = -f(x)$.
    (b) Even; $g(-x) = 5(-x)^4-2(-x)^2+11 = 5x^4-2x^2+11 = g(x)$.
    (c) Neither; $h(-x) = 3(-x)^3+5(-x)-4 = -3x^3-5x-4$ which is neither $h(x)$
       nor $-h(x)$.
    (d) Neither; $F(-x) = (-x)^6-2(-x)^4+2(-x)-5 = x^6-2x^4-2x-5$ which is
       neither $F(x)$ nor $-F(x)$.
    (e) Even; $G(-x) = 4/[(-x)^2+5] = 4/(x^2+5) = G(x)$.
    (f) Odd; $H(-x) = [(-x)^2+5]/[2(-x)^3+(-x)] = (x^2+5)/[-(2x^3+x)] = -H(x)$.

6.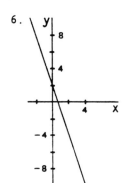

9.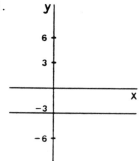

12.

15.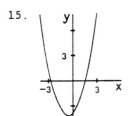

18.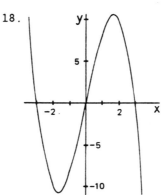

21.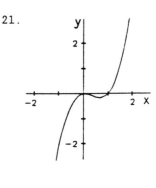

24.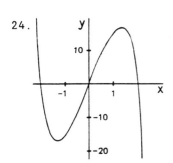

27.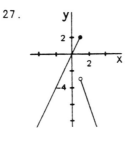

30.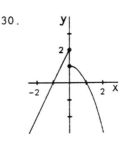

33.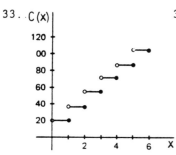

36. (a)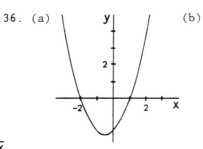

(b)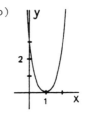

36. (c)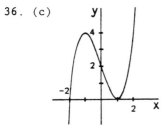

(d)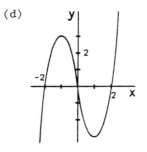

(e)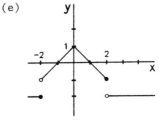

39. $R(x) = \begin{cases} 160x, & x = 1,2,3,\cdots,30 \\ [160 - 2(x-30)]x, & x = 31,32,33,\cdots,70 \end{cases}$

$= \begin{cases} 160x, & x = 1,2,3,\cdots,30 \\ 2x(110-x), & x = 31,32,33,\cdots,70 \end{cases}$; $R(35) = 2(35)(110-35) = \$5,250$

$R(70) = 2(70)(110-70) = \$5,600$

Problem Set 1.3   Linear Functions: Constant Slope

3. $m = \dfrac{0-2}{3-(-4)} = \dfrac{-2}{7}$

6. $m = \dfrac{\sqrt{2} - (\sqrt{3} + \sqrt{5})}{1.642 - \pi} \approx 1.703$

9. $m = \dfrac{9-1}{5-3} = 4;\ y-1 = 4(x-3).$

12. $m = \dfrac{15-3}{-10-2} = -1;\ y-3 = -(x-2).$

15. Change $5x-y = 12$ to $y = 5x-12$ to see that the slope is 5. The line we seek is $y-(-1) = 5(x-2)$ or $5x-y = 11.$

18. The slope of $y = 3x-5$ is 3, so slope of perpendicular line is $-\frac{1}{3}$. Then the line we seek is $y-1 = -\frac{1}{3}(x-4)$ or $x+3y = 7.$

21. The point of intersection is $(3,-2)$, so the line we seek is $y+2 = \frac{3}{4}(x-3)$ or $3x-4y = 17.$

24. The point of intersection is $(4,-2)$ and first line has slope $-\frac{1}{2}$, so the line we seek has slope 2 and equation $y+2 = 2(x-4)$ or $2x-y = 10.$

27. Two points of the line are $(-2,4)$ and $(3,2)$, so its slope is -0.4 Then an equation of the line is $y-4 = -0.4(x+2)$ or $y = -0.4x + 3.2$ Expressing this as $f(x) = -0.4x + 3.2$ gives $f(5) = -0.4(5) + 3.2 = 1.2$

30. Points of the line are $(0,2)$ and $(3,0)$, so the slope is $-\frac{2}{3}$. Thus, the slope-intercept equation of the line is $y = -\frac{2}{3}x + 2.$

33. Points of the line are $(0,20000)$ and $(10,2000)$; the slope is -1800 and y-intercept is 20000. $y = -1800x + 20000$ or $f(x) = -1800x + 20000.$

36. (a) Slope of $y = -2x+4$ is -2.
    An equation of line through $(7,5)$ and perpendicular to the given line is $y-5 = \frac{1}{2}(x-7)$ or $x-2y = -3.$

    (b) Point of intersection of $2x+y = 4$ and $x-2y = -3$ is $(1,2).$

    (c) Distance between $(7,5)$ and $y = -2x+4$ is the distance between $(7,5)$ and $(1,2)$, $\sqrt{(1-7)^2 + (2-5)^2} = \sqrt{45} \approx 6.708$

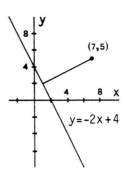

Problem Set 1.4   Applications of Linear Functions

3. (a) $S(x) = 0.04x + 1200$ (b) $S(36000) = 0.04(36000) + 1200 = \$2,640$

6. $A(x) = 10000 + 0.09x$ 9. $C(x) = 42000 + 250x$

12. $R(x) = 475x$.   Break even if $475x = 300x + 2800$; i.e., $x = 160$ freezers.

15. $s(t) = 480 - 60t$, t in $[0,8]$ 18. (a) $16/5 = 3.2$ R-rating per inch,
                                                  so $R(t) = 4 + 3.2t$.
                                              (b) If $40 = 4 + 3.2t$, $t = 11.25$,
                                                  so 11.25 inches is needed.

21. (a) Rate of change of F with respect to C is the slope, $\frac{9}{5}$.

    (b) $C = \frac{5}{9}(F-32)$; rate of change of C with respect to F is $\frac{5}{9}$.

24. (a) $\dfrac{17.4 - 12}{4 - 0} = 1.35$ is the slope, so $P(t) = 1.35t + 12$.

    (b) Rate of change of $P(t)$ with respect to t is 1.35% per year.

Problem Set 1.5   Nonlinear Functions: Variable Slope

3. Tangent at $(4,2)$ goes through $(1,3)$, so the slope is $\frac{1}{-3} = -\frac{1}{3}$.

6. Tangent at $(5,3)$ goes through $(3,8)$, so slope is $\frac{5}{-2} = -2.5$.

9. (a) at A & E   (b) at C   (c) at B & D

12. 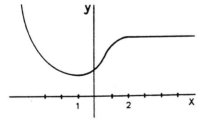   15. Slope of $y = x^3$ at x is $3x^2$, so
                                    slope at $x = 4$ is $3(4)^2 = 48$.

                                18. Slope of $y = x^8$ at x is $8x^7$, so
                                    slope at $x = 2$ is $8(2)^7 = 1024$.

21. $\dfrac{dy}{dx} = \dfrac{d}{dx}(2x^4) - \dfrac{d}{dx}(3x^3) + \dfrac{d}{dx}(5x^2) + \dfrac{d}{dx}(\pi)$

$\quad = 2 \cdot 4x^3 - 3 \cdot 3x^2 + 5 \cdot 2x + 0 = 8x^3 - 9x^2 + 10x.$

24. $\dfrac{dy}{dx} = \dfrac{d}{dx}(\tfrac{3}{4}x^{12}) - \dfrac{d}{dx}[(\tfrac{1}{3})x^6)] + \dfrac{d}{dx}(\tfrac{5}{7})$

$\quad = \tfrac{3}{4} \cdot 12x^{11} - \tfrac{1}{3} \cdot 6x^5 + 0 = 9x^{11} - 2x^5$

27. $\dfrac{dy}{dx} = 6x^5 - 36x^2 + 5$, so the slope at x=2 is $6(2)^5 - 36(2)^2 + 5 = 53$. $y = -30$ when x = 2, so an equation of the tangent is $(y+30) = 53(x-2)$.

30. $\dfrac{dy}{dx} = 6x^2 - 10$ is the slope at x. The slope is 44 if $6x^2 - 10 = 44$, or x is 3 or -3. The corresponding y values on the curve are 2 and -46, so the points at which the slope is 44 are (3,2) and (-3,-46).

33. $\dfrac{dy}{dx} = 3x^2 - 9$ is the slope at x. The slope is 3 if $3x^2 - 9 = 3$, or x is 2 or -2. The corresponding y values on the curve are 0 and 20, so the points are (2,0) and (-2,20). The corresponding tangent lines are $y = 3(x-2)$ and $y-20 = 3(x+2)$.

36. $\dfrac{dy}{dx} = 4x+4$ is the slope at x. It is positive if $x > -1$.

39. $\dfrac{dy}{dx} = 3\pi x^2 - 2\sqrt{2}x$, so the slope at x = 1.592 is

$$3\pi(1.592)^2 - 2\sqrt{2}(1.592) \approx 19.384.$$

Problem Set 1.6   Variable Rates of Change

3. Let $V = 100t^3 - 25t^2 + 50$, so $\dfrac{dV}{dt} = 300t^2 - 50t$. At the end of 2 minutes water is entering at the rate of $300(2)^2 - 50(2) = 1100$ gallons/minute.

6. (a) (2,2.8) is the point of tangency. The tangent line also goes through (6,4.8), so its slope, the rate of change of y with respect to x, is $\dfrac{4.8-2.8}{6-2} = \dfrac{1}{2}$.

6. (b) $(4,2.5)$ is the point of tangency. The tangent line also goes through $(0,4.5)$, so its slope, the rate of change of y with respect to x, is $\frac{-2}{-4} = -\frac{1}{2}$.

9. (a) $\frac{(3.1)^2-(3)^2}{3.1-1} = 6.1$ ft/sec   (b) $\frac{(3.01)^2-(3)^2}{3.01-1} = 6.01$ ft/sec

   (c) $\frac{(3.001)^2-(3)^2}{3.001-1} = 6.001$ ft/sec   (d) It seems that the velocity when $t=3$ is 6 ft/sec.

12. (a) $v(t) = ds/dt = -32t-96$, so $v(2) = -32(2) - 96 = -160$ ft/sec and $v(4) = -32(4) - 96 = -224$ ft/sec.

   (b) It occurs at the end of 4 seconds. [See part (a).]

   (c) $s(10) = -16(10)^2 - 96(10) + 3000 = 440$ ft.
       $v(10) = -32(10) - 96 = -416$ ft/sec.

15. (a) The points involved are $(10,13)$ and $(20,27)$ so the average rate of increase is $\frac{27-13}{20-10} = 1.4$ thousand persons per year.

   (b) The graph is steepest at $t = 15$, so rate of increase was greatest in 1965.

   (c) Maximum rate of increase is slope of graph at $t = 15$. Two points of the tangent line seem to be $(10,10)$ and $(15,20)$, so the slope is $\frac{10}{5} = 2$ thousand persons per year.

18. (a) Marginal cost is $dC/dx = -0.1x + 250$.
       When $x = 500$, marginal cost is $-0.1(500) + 250 = \$200$ per chair.

   (b) The cost of making the 501st chair is about $200. (marginal cost)

21. (a) About $8,000.   (b) About 38% (slope of tangent is about 0.38).

24. (a) Ball is at maximum height when $v = 0$. $v = \frac{ds}{dt} = -32t+48$ which is zero when $t = 1.5$ seconds.

   (b) $s(1.5) = -16(1.5)^2 + 48(1.5) + 160 = 196$ feet.

   (c) Ball hit the ground when $s = 0$. $0 = -16t^2+48t+160 = -16(t-5)(t+2)$ when $t = 5$ seconds.

   (d) $v(5) = -32(5) + 48 = -112$ ft/sec, so the speed was 112 ft/sec.

Problem Set 1.6   7

27. (a) N(0) = 25,000 bacteria.
    (b) $\frac{dN}{dt}$ = 1400 - 1000t; when t = $\frac{1}{2}$, $\frac{dN}{dt}$ = 1400 - 1000($\frac{1}{2}$) = 900 bacteria per hour.
    (c) $\frac{dN}{dt}$ = 0 when t = 1.4 hours, so population begins to decrease after 1.4 hours (1 hour, 24 minutes).

Chapter 1 Review Problem Set

3. (a) Neither since f(-x) ≠ f(x) and f(-x) ≠ -f(x).
   (b) Odd since g(-x) = -g(x).   (c) Even since h(-x) = h(x).

6. (a) 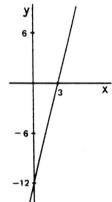   (b) 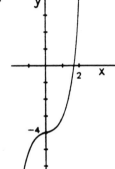   (c)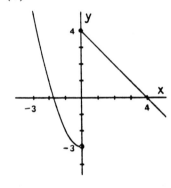

9. (a) Slope 3/2; y-intercept is -6.   (b) Slope 3; y-intercept is 5.
   (c) Slope -4/3; y-intercept is 4.

12. Slope of 3x-y = 8 is 3.   Then the line we seek is y-4 = 3(x+2).

15. $\frac{dy}{dx}$ = 10x⁴-30x²+6x, so when x = 1 the slope is -14; and y = -7. Thus, an equation of the tangent is y+7 = -14(x-1).

18. F(x) = 320000 + 2000x

21. Slope of the line is $\frac{90,000 - 2,000}{0 - 8}$ = -11,000 dollars per year.

   f(t) = -11000t + 90000

24.

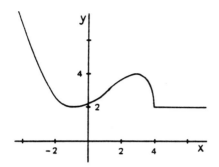

27. (a) $\dfrac{50-20}{7-2} = 6$ percent per year.

   (b) Slope of tangent at $(7,50)$ is about 10 percent per year.

30. (a) $C'(x) = 46 - 0.02x$, so $C'(1200) = 46 - 0.02(1200) = 22$.
     Therefore, marginal cost when 1200 pairs are made is $22.00/pair.

     $R'(x) = 70 - 0.036x$, so $R'(1200) = 70 - 0.036(1200) = 26.8$.
     Thus, marginal revenue when 1200 pairs are made is $26.80/pair.

     $P(x) = R(x) - C(x)$, so $P'(x) = R'(x) - C'(x)$. Therefore, marginal
     profit when 1200 pairs are made is $26.80 - $22.00 = $4.80/pair.

CHAPTER 2   THE DERIVATIVE

Problem Set 2.1   The Limit Concept

3. Exists and equals 2.

6. Doesn't exist.

9. (a)

(b) $\lim\limits_{x \to 3} f(x) = 9$

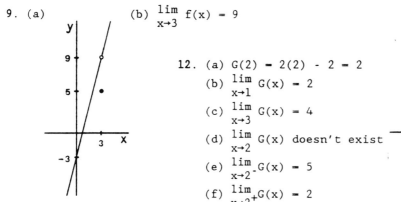

12. (a) $G(2) = 2(2) - 2 = 2$
    (b) $\lim\limits_{x \to 1} G(x) = 2$
    (c) $\lim\limits_{x \to 3} G(x) = 4$
    (d) $\lim\limits_{x \to 2} G(x)$ doesn't exist
    (e) $\lim\limits_{x \to 2^-} G(x) = 5$
    (f) $\lim\limits_{x \to 2^+} G(x) = 2$

15. $\lim\limits_{x \to -3} \dfrac{(x+3)(x-2)}{x+3} = \lim\limits_{x \to -3} (x-2)$

$= -5$

18. $\lim\limits_{x \to 5} \dfrac{(2x+3)(x-5)}{x-5} = \lim\limits_{x \to 5} (2x+3)$

$= 13$

21. $\lim\limits_{x \to 5} \dfrac{x(x-5)}{(x+3)(x-5)} = \lim\limits_{x \to 5} \dfrac{x}{x+3}$

$= \dfrac{5}{8}$

24. $\lim\limits_{x \to 0} \dfrac{x^2(x-1)}{x(x+2)} = \lim\limits_{x \to 0} \dfrac{x(x-1)}{x+2} = 0$

27. It equals $\dfrac{(2-1)(4+4+4)}{4+10-6} = \dfrac{3}{2}$

30. $\lim\limits_{x \to 3}(2x^4-7x^2+9x-1) \overset{[4]}{=} \lim\limits_{x \to 3}(2x^4) - \lim\limits_{x \to 3}(7x^2) + \lim\limits_{x \to 3}(9x) - \lim\limits_{x \to 3}(1)$

$\overset{[3,1]}{=} 2\lim\limits_{x \to 3}(x^4) - 7\lim\limits_{x \to 3}(x^2) + 9\lim\limits_{x \to 3}(x) - 1$

$\overset{[7,2]}{=} 2\left(\lim\limits_{x \to 3} x\right)^4 - 7\left(\lim\limits_{x \to 3} x\right)^2 + 9(3) - 1 \overset{[2]}{=} 2(3)^4-7(3)^2+27-1 = 125$

33. $\dfrac{(5)^2+2(5)+5}{(5)+3} = \dfrac{40}{8} = 5$

36. $[2(6)^2-10(6)-9]^4 = 81$

39. $\sqrt{(2)^2+5} + 2\sqrt[3]{2(2)^3+11} = 3+6 = 9$

42. $\dfrac{(4+2+1)(4-1)(2+3)}{(6-5)^4(4+1)} = \dfrac{105}{5} = 21$

45. (a) $(5-2-1)^4 = 16$

(b) $\left(\dfrac{5-4}{-4+6}\right)^4 = \dfrac{1}{16}$

(c) $\sqrt[3]{7-5(-4)} = 3$

(d) $[8+4+2+1-2(5)]^3 = 125$

48. $(-8+10)\sqrt{5-(-20)} = (2)(5) = 10$

51. $\lim\limits_{x\to 2} \dfrac{2(x-2)(x^2+2x+4)}{3(x-2)(x+2)} = \lim\limits_{x\to 2} \dfrac{2(x^2+2x+4)}{3(x+2)} = \dfrac{2(4+4+4)}{3(2+2)} = 2$

54. Doesn't exist; denominator approaches 0 while numerator approaches 13.

57. $\lim\limits_{x\to 1^-}F(x) = \lim\limits_{x\to 1^-}(2x+1) = 3;\ \lim\limits_{x\to 1^+}F(x) = \lim\limits_{x\to 1^+}(3x) = 3.$

Therefore, $\lim\limits_{x\to 1} F(x) = 3$

60.

x	$\dfrac{\ln(x+1)}{x^2+2x}$	$\dfrac{\sqrt{2+x}-\sqrt{2}}{x}$
0.1	0.4538580	0.3492411
0.01	0.4950413	0.3531125
0.001	0.4995004	0.3535092
-0.001		0.3535976

(a) It seems that $\lim\limits_{x\to 0^+} \dfrac{\ln(x+1)}{x^2+2x} = 0.5$

(b) What $\lim\limits_{x\to 0} \dfrac{\sqrt{2+x}-\sqrt{2}}{x}$ is, is not clear but it seems that it might be a number which begins 0.35. In fact, it is $\sqrt{2}/4 \approx 0.35355339$.

Problem Set 2.2 The Derivative

3. $\lim\limits_{h\to 0} \dfrac{(x+h)^3 - x^3}{h} = \lim\limits_{h\to 0} \dfrac{(x^3+3x^2h+3xh^2+h^3) - x^3}{h} = \lim\limits_{h\to 0} \dfrac{h(3x^2+3xh+h^2)}{h}$

$= \lim\limits_{h\to 0}(3x^2+3xh+h^2) = 3x^2 + 0 + 0 = 3x^2$

6. $\displaystyle\lim_{h\to0}\frac{\dfrac{1}{x+h+1}-\dfrac{1}{x+1}}{h} = \lim_{h\to0}\frac{(x+1)-(x+h+1)}{h(x+h+1)(x+1)} = \lim_{h\to0}\frac{-h}{h(x+h+1)(x+1)}$

$$= \lim_{h\to0}\frac{-1}{(x+h+1)(x+1)} = \frac{-1}{(x+1)^2}$$

9. $\displaystyle\lim_{h\to0}\frac{\sqrt{x+h+2}-\sqrt{x+2}}{h} = \lim_{h\to0}\frac{(x+h+2)-(x+2)}{h(\sqrt{x+h+2}+\sqrt{x+2})} = \lim_{h\to0}\frac{h}{h(\sqrt{x+h+2}+\sqrt{x+2})}$

$$= \lim_{h\to0}\frac{1}{(\sqrt{x+h+2}+\sqrt{x+2})} = \frac{1}{2\sqrt{x+2}}$$

12. $f(x) = x^{3/2}$; $f'(x) = \frac{3}{2}x^{1/2}$ **15.** $f(x) = 6x^{1/3}$; $f'(x) = 2x^{-2/3}$

18. $f'(x) = 2x^{\sqrt{2}-1}$ **21.** $f'(x) = -11\sqrt{3}x^{\sqrt{3}-1}$

24. (a) $(x^4)^{3/2}/9^{3/2} = x^6/27$ (b) $x^2x^{1/4} = x^{9/4}$

    (c) $x^3/x^{1/2} = x^{5/2}$ (d) $x^{1/2}x^2x^{-2/3} = x^{3/6}x^{12/6}x^{-4/6} = x^{11/6}$

    (e) $x^{3/2}9^{-3/2}x^{-3/2}x^2 = x^2/9^{3/2} = x^2/27$

    (f) $x^{2/3} + 2x^{1/3}x^{5/3} + x^{10/3} = x^{2/3} + 2x^2 + x^{10/3}$

27. (a) $\dfrac{dy}{dx} = 0$ (b) $\dfrac{dy}{dx} = 28x^6$ (c) $\dfrac{dy}{dx} = 31x^{2.1}$

    (d) $\dfrac{dy}{dx} = 14x^{1/6}$ (e) $\dfrac{dy}{dx} = -3(2\pi+1)x^{2\pi}$ (f) $\dfrac{dy}{dx} = 0$ $[\pi^{3.1}$ is a constant.$]$

30. $\displaystyle\lim_{h\to0}\frac{\sqrt[3]{x+h}-\sqrt[3]{x}}{h} = \lim_{h\to0}\frac{(x+h)-x}{h[\sqrt[3]{(x+h)^2}+\sqrt[3]{x(x+h)}+\sqrt[3]{x^2}]}$

$$= \lim_{h\to0}\frac{h}{h[\sqrt[3]{(x+h)^2}+\sqrt[3]{x(x+h)}+\sqrt[3]{x^2}]}$$

$$= \lim_{h\to0}\frac{1}{[\sqrt[3]{(x+h)^2}+\sqrt[3]{x(x+h)}+\sqrt[3]{x^2}]} = \lim_{h\to0}\frac{1}{3\sqrt[3]{x^2}}$$

33. $F(r) = 400r^{-2}$; $F'(r) = -800r^{-3}$; $F'(10) = -800(10)^{-3} = -0.8$ dynes/cm.

36. Point: When $x = \frac{1}{4}$, $y = 4(\frac{1}{4})^{3/2} = 4(\frac{1}{8}) = \frac{1}{2}$.

Slope: $dy/dx = 6x^{1/2}$, which equals $6(\frac{1}{4})^{1/2} = 3$, when $x = \frac{1}{4}$.

Equation: $y - \frac{1}{2} = 3(x - \frac{1}{4})$, or $12x-4y = 1$.

Problem Set 2.3   Rules for Differentiation

3. $f(x) = 2x^3 + 5x^{1/2} - 4x^{-5}$; $f'(x) = 6x^2 + \frac{5}{2}x^{-1/2} + 20x^{-6}$

6. $f(\dot{x}) = x^{4/3} - 2x^{-1/3}$; $f'(x) = \frac{4}{3}x^{1/3} + \frac{2}{3}x^{-4/3}$

9. Point: When $x = -1$, $y = 2(-1)^3 - 4(-1)^2 - 3/(-1) = -3$.

Slope: $dy/dx = 6x^2-8x+3x^{-2}$, which equals 17, when $x = -1$.

Equation: $y+3 = 17(x+1)$, or $y = 17x+14$.

12. Point: When $x = 8$, $y = 3(8)^{2/3} - 8/(8)^{1/3} = 12-4 = 8$.

Slope: $dy/dx = 2x^{-1/3} + \frac{8}{3}x^{-4/3}$, which equals $\frac{7}{6}$, when $x = 8$.

Equation: $y-8 = \frac{7}{6}(x-8)$, or $7x-6y = 8$.

15. $F'(x) = (x^2+5)(6x^2-3) + (2x^3-3x+9)(2x) = 10x^4+21x^2+18x-15$

18. $F'(x) = (3+5x^{-1})(8x) + (7+4x^2)(-5x^{-2}) = 24x+20-35x^{-2}$

21. $f'(t) = (t-16t^{-1})[\frac{3}{2}t^{-1/2} - \frac{3}{2}t^{1/2}] + (3t^{1/2}-t^{3/2})(1+16t^{-2})$

$f'(4) = (4-4)[\frac{3}{4} - 3] + (6-8)(1+1) = -4$

24. (a) $F'(x) = \dfrac{(x^{1/2})(3) - (3x-2)[(1/2)x^{-1/2}]}{x} = \dfrac{3x^{1/2} - (3/2)x^{1/2} + x^{-1/2}}{x}$

$= \dfrac{(3/2)x^{1/2} + x^{-1/2}}{x} = (3/2)x^{-1/2} + x^{-3/2}$

(b) $F(x) = 3x^{1/2} - 2x^{-1/2}$; $F'(x) = \frac{3}{2}x^{-1/2} + x^{-3/2}$

27. $f(x) = \dfrac{4x^{1/2}}{3+x^{1/2}};$ $f'(x) = \dfrac{(3+x^{1/2})(2x^{-1/2}) - (4x^{1/2})[(1/2)x^{-1/2}]}{(3+x^{1/2})^2}$

$$= \dfrac{6x^{-1/2} + 2 - 2}{(3+x^{1/2})^2} = \dfrac{6x^{-1/2}}{(3+x^{1/2})^2}$$

30. $dy/dx = (3x^2+x-7)(-12x^2) + (5-4x^3)(6x+1) = -60x^4-16x^3+84x^2+30x+5$

33. $\dfrac{dy}{dx} = 4x + 3 - \dfrac{(x-4)(1) - (x)(1)}{(x-4)^2} = 4x + 3 + 4(x-4)^{-2}$

36. $H'(x) = G(x)G'(x) + G(x)G'(x) = 2G(x)G'(x)$

39. $v(t) = s'(t) = \dfrac{(t+1)[4t - (3/2)t^{-1/2}] - (2t^2-3t^{1/2})(1)}{(t+1)^2}$

$v(4) = \dfrac{(5)(16 - 3/4) - (32-6)}{25} = 2.01$ cm/sec

42. (a) $C(500) = 2000 + 12(500) - 8\sqrt{500} + 0.05\sqrt{(500)^3} = \$8,380.13$

(b) Average cost $= 8,380.13/500 = \$16.76$

(c) $C'(x) = 12 - 4x^{-1/2} + 0.075x^{1/2}$

$C'(500) = 12 - 4(500)^{-1/2} + 0.075(500)^{1/2} = \$13.50$ per jacket

(d) Let $A(x) = C(x)/x = 2000x^{-1} + 12 - 8x^{-1/2} + 0.05x^{1/2}$ be avg. cost.

$A'(x) = -2000x^{-2} + 4x^{-3/2} + 0.025x^{-1/2}$

$A'(500) = -2000(500)^{-2} + 4(500)^{-3/2} + 0.025(500)^{-1/2}$

$\approx -\$0.006.$ [That is, the production of the 501st jacket decreases the average cost per jacket by less than 1 cent.]

Problem Set 2.4 The General Power Rule

3. $f'(x) = 8(3x^2+x+1)^7(6x+1)$ 6. $f'(x) = 6(x^{2/3}-2x)^5[(\tfrac{2}{3}x^{-1/3}-2]$

9. $f'(x) = \tfrac{1}{2}(2+3x-x^2)^{-1/2}(3-2x)$

12. $f'(x) = \frac{3}{4}(x^4-2x^3+2)^{-1/4}(4x^3-6x^2)$

15. $f(x) = 4(3x-1)^{-5}$; $f'(x) = -20(3x-1)^{-6}(3) = -60(3x-1)^{-6}$

18. $f(x) = 4(2x+3)^{-3/2}$; $f'(x) = -12(2x+3)^{-5/2}$

21. $\dfrac{dy}{dx} = 3\left(\dfrac{2x-1}{3x+2}\right)^2 \dfrac{(3x+2)(2) - (2x-1)(3)}{(3x+2)^2} = \dfrac{21(2x-1)^2}{(3x+2)^4}$

24. $\dfrac{dy}{dx} = 3\left(\dfrac{3x-1}{x^2+x-4}\right)^2 \dfrac{(x^2+x-4)(3) - (3x-1)(2x+1)}{(x^2+x-4)^2} = \dfrac{3(3x-1)^2(-3x^2+2x-11)}{(x^2+x-4)^4}$

27. $dy/dx = (3x^2)[\frac{5}{3}(2x^2+3)^{2/3}(4x)] + (2x^2+3)^{5/3}(6x)$

$\qquad = 2x(2x^2+3)^{2/3}(16x^2+9)$

30. $dy/dx = (2x^3+1)^4[3(x^3+x)^2(3x^2+1)] + (x^3+x)^3[4(2x^3+1)^3(6x^2)]$

$\qquad = 3(2x^3+1)^3(x^3+x)^2(14x^5+10x^3+3x^2+1)$

33. $F(x) = (2x-1)^8(6x^{1/3}+1)$
$\qquad F'(x) = (2x-1)^8(2x^{-2/3}) + (6x^{1/3}+1)[8(2x-1)^7(2)]$
$\qquad F'(1) = (1)(2) + (7)[8(1)(2)] = 114$

36. **Point:** When $x = 3$, $y = (12)(1) = 12$.

   Slope: $\dfrac{dy}{dx} = (x^2+3)[-\frac{2}{5}(x^2+2x-14)^{-7/5}(2x+2)] + (x^2+2x-14)^{-2/5}(2x)$.

   Thus, when $x=3$, slope is $(12)[-\frac{2}{5}(1)(8)] + (1)(6) = -32.4$

   Equation: $y-12 = -32.4(x-3)$, or $162x+5y = 546$.

39. $F(x) = (4x-10-8x^{-1})^{3/2}$; $F'(x) = \frac{3}{2}(4x-10-8x^{-1})^{1/2}(4+8x^{-2})$

$\qquad F'(4) = 1.5(16-10-2)^{1/2}(4+0.5) = 13.5$

42. $F(x) = (2x-7)^5(2x+1)^{-1/2}$

$\qquad F'(x) = (2x-7)^5[-\frac{1}{2}(2x+1)^{-3/2}(2)] + (2x+1)^{-1/2}[5(2x-7)^4(2)]$

$\qquad F'(4) = (8-7)^5[-\frac{1}{2}(8+1)^{-3/2}(2)] + (8+1)^{-1/2}[5(8-7)^4(2)] = \frac{89}{27}$

45. (a) $dA/dr = 40000(1 + r/2)^7(1/2) = 20000(1 + r/2)^7$

(b) When $r = 0.12$, $A = 5000(1.06)^8 = \$7,969.24$ and
$dA/dr = 20000(1.06)^7 = \$30,072.61$ per unit (100%) which is a bit
misleading; of more practical interest is the rate of change of
A per 0.01 unit (1%), which is \$300.73.

48. (a) $I(10) = 4000/100 + 1000/100 = 50$
$I(15) = 4000/225 + 1000/25 \approx 57.78$

(b) $I'(x) = -8000x^{-3} - 2000(20-x)^{-3}(-1) = -8000x^{-3} + 2000(20-x)^{-3}$
$I'(10) = -8000/1000 + 2000/1000 = -6$
$I'(15) = -8000/3375 + 2000/125 \approx 13.63$

(c) As P is moved toward S_2 from $x = 15$, x is increasing so I will
increase since $I'(15)$ is positive.

Problem Set 2.5   Higher Order Derivatives

3. $dy/dx = -4x^{-2}$; $d^2y/dx^2 = 8x^{-3}$; $d^3y/dx^3 = -24x^{-4}$

6. $dy/dx = \frac{3}{4}x^4 - 6x^{-3}$; $d^2y/dx^2 = 3x^3 + 18x^{-4}$; $d^3y/dx^3 = 9x^2 - 72x^{-5}$

9. $h(s) = 3(s+4)^{-1/2}$; $h'(s) = -\frac{3}{2}(s+4)^{-3/2}$; $h''(s) = \frac{9}{4}(s+4)^{-5/2}$

$h'''(s) = -\frac{45}{8}(s+4)^{-7/2}$; thus, $h'''(0) = -\frac{45}{8}(4)^{-7/2} = -\frac{45}{1024}$

12. $dy/dx = 2(x^3-x+2)(3x^2-1)$

$d^2y/dx^2 = [2(x^3-x+2)](6x) + (3x^2-1)[2(3x^2-1)] = 30x^4 - 24x^2 + 24x + 2$

15. $s_0 = 256$ and $v_0 = 96$, so $s(t) = -16t^2 + 96t + 256$, $v(t) = s'(t) = -32t+96$,
and $a(t) = v'(t) = -32$ are the appropriate equations of motion.

(a) Given above
(b) $s(2) = -16(4) + 96(2) + 256 = 384$ ft
(c) $v(2) = -32(2) + 96 = 32$ ft/sec
(d) Max. height when $0 = v = -32t+96$; that is, when $t = 3$ seconds
(e) Hit ground when $0 = s = -16t^2 + 96t + 256 = -16(t-8)(t+2)$; $t = 8$ sec
(f) $v(8) = -32(8) + 96 = -160$ ft/sec
(g) $a(2) = -32$ ft/sec^2 and $a(6) = -32$ ft/sec^2

18. Eqtns. of motion: $s(t) = 3t-12t^{1/2}+2$, $v(t) = 3-6t^{-1/2}$; $a(t) = 3t^{-3/2}$

(a) $s(1) = 3-12+2 = -7$ m; $v(1) = 3-6 = -3$ m/min; $a(1) = 3$ m/min^2

(b) $0 = v = 3-6t^{-1/2}$; therefore t=4, and $s(4) = 12-24+2 = -10$ meters.

(c) v approaches 3 m/min as t gets larger since $6t^{-1/2}$ approaches 0.
(d) a approaches 0 m/min^2 as t gets larger.

21. (a) $dI/dt = 0$ during all of the last year, but dI/dt and d^2I/dt^2 are both expected to be positive during the years ahead.
(b) dP/dt is very negative at the present time, but d^2P/dt^2 is expected to become positive soon and then dP/dt is expected to become positive in about two years.
(c) dT/dt is still positive but the penicillin seems to be causing d^2T/dt^2 to be negative.
(d) d^2s/dt^2 is positive now because the accelerator is pressed to the floor but after another two minutes that won't do any good.

24. $f(x) = 12x^{-2/3}+x^{1/2}$; $f'(x) = -8x^{-5/3} + \frac{1}{2} x^{-1/2}$

$f''(x) = \frac{40}{3} x^{-8/3} - \frac{1}{4} x^{-3/2}$; $f'''(x) = -\frac{320}{9} x^{-11/3} + \frac{3}{8} x^{-5/2}$

Therefore, $f'''(1) = -\frac{320}{9} + \frac{3}{8} = -\frac{2533}{72} \approx 35.18$

27. $dy/dx = 3(4+x)^2$ is the slope function. The rate of change of the slope is $d^2y/dx^2 = 6(4+x)$, which equals 30 when x=1.

30. (a) The rate at which the rumor is spreading is dN/dt, which is the slope of the tangent line. It is positive for t in (0,7).
(b) The rate is decreasing when the slope is decreasing. That is occurring for t about in (4,7).
(c) The rumor is spreading most rapidly when the slope is greatest. That occurs approximately when t = 3.

Chapter 2 Review Problem Set

3. $3(-2)+8 = 2$

6. $(-3-2)^2 + \sqrt{3} = 25 + \sqrt{3}$

9. $\lim_{x \to 5} \frac{2(x-5)}{x(x-5)} = \lim_{x \to 5} \frac{2}{x} = \frac{2}{5}$

12. $\lim_{x \to 5} \frac{(x+2)(x-5)}{(x+5)(x-5)} = \lim_{x \to 5} \frac{x+2}{x+5} = \frac{7}{10}$

15. $\displaystyle\lim_{x\to 5}\frac{(x+2)(x-3)}{(x+2)(x+5)} = \lim_{x\to 5}\frac{x-3}{x+5} = \frac{-5}{3}$

18. $\displaystyle\lim_{x\to 5}\frac{(x-5)(x^2+5x+25)}{(x-5)(x+3)} = \lim_{x\to 5}\frac{x^2+5x+25}{x+5} = \frac{75}{8}$

21. $\dfrac{2^3}{3} = \dfrac{8}{3}$

24. $[2(2)^3-2(2)-7]\sqrt{5(2)^2+7} = 5\sqrt{27} = 15\sqrt{3}$

27. $\displaystyle\lim_{x\to 3}\frac{(2x-5)^6(x-3)}{(x+3)(x-3)(x^2+x-11)} = \lim_{x\to 3}\frac{(2x-5)^6}{(x+3)(x^2+x-11)} = \frac{1}{6}$

30. $\dfrac{10}{\sqrt{25}} - 3\sqrt[3]{-8} = 2-3(-2) = 8$

33. $\displaystyle\lim_{h\to 0}\frac{\dfrac{2}{\sqrt{x+h}} - \dfrac{2}{\sqrt{x}}}{h} = \lim_{h\to 0}\frac{-2}{\sqrt{x(x+h)}(\sqrt{x} + \sqrt{x+h})} = \frac{-1}{x\sqrt{x}}$

36. $f(x) = 8x^{1/4};\ f'(x) = 8(\tfrac{1}{4})x^{-3/4} = 2x^{-3/4}$

39. $f(x) = 2x^{3/2};\ f'(x) = 3x^{1/2}$

42. $f(x) = (x^4x^{1/2})^{1/3} = (x^{9/2})^{1/3} = x^{3/2};\ f'(x) = \tfrac{3}{2}x^{1/2}$

45. $f(x) = (x^2-2x+5)^{1/2};\ f'(x) = (x^2-2x+5)^{-1/2}(x-1)$

48. $f(x) = (2x+4x^{-1})^4;\ f'(x) = 4(2x+4x^{-1})^3(2-4x^{-2}) = 64x^{-5}(x^2+2)^3(x^2-2)$

51. $f'(x) = 3\left(\dfrac{2x^2+1}{x^3-2x+4}\right)^2 \dfrac{(x^3-2x+4)(4x) - (2x^2+1)(3x^2-2)}{(x^3-2x+4)^2}$

which simplifies to $\dfrac{3(2x^2+1)^2(-2x^4-7x^2+16x+2)}{(x^3-2x+4)^4}$.

54. $f(x) = (2x^2+x+1)(4x+1)^{1/2}$

   $f'(x) = (4x+1)(4x+1)^{1/2} + (2x^2+x+1)[\frac{1}{2}(4x+1)^{-1/2}(4)]$

   which can be simplified to $(4x+1)^{-1/2}(20x^2+10x+3)$.

57. $f'(x) = 5[(x^2+2x)^4+3]^4[4(x^2+2x)^3(2x+2)] = 40(x+1)(x^2+2x)^3[(x^2+2x)^4+3]^4$

60. $h'(s) = 36(2s+1)^{-2/3}$; $h''(s) = -48(2s+1)^{-5/3}$; $h'''(s) = 160(2s+1)^{-8/3}$.

   Therefore, $h'''(-1) = 160(-1)^{-8/3} = 160$.

63. The slope of $x-2y = 5$ is $\frac{1}{2}$. For $y = 4x^{1/2}$, $y'(x) = 2x^{-1/2}$ which is $\frac{1}{2}$

   if x is 16. Then, y is 16 if x is 16, so the point is (16,16).

66. $F(t) = 72-15t+6(t+2)^{-2}$

   $F'(t) = -15-12(t+2)^{-3}$, so $F'(2) = -15-12(4)^{-3} = -15.1875$.

   Thus, the temperature is dropping about 15 °F per hour when t is 2.

69. (a) When $0 \leq t \leq 17$ (when graph is rising).

   (b) When $0 \leq t < 10$ (when graph is concave up).

   (c) When $t = 10$ (when graph is steepest).

Problem Set 3.1   Continuity and Differentiability

3. Not continuous   6. Not continuous

9. $\lim\limits_{x \to 1} f(x) = \lim\limits_{x \to 1} \dfrac{x^2-4x}{x-4} = \dfrac{-3}{-3} = 1$, and $f(1) = 1$, so f is continuous at 1.

$\lim\limits_{x \to 4} f(x) \neq f(4)$ since $f(4)$ doesn't exist, so f is discontinuous at 4.

12. $\lim\limits_{x \to 10} f(x) = \lim\limits_{x \to 10} \dfrac{2x^2-10x}{3x-15} = \dfrac{100}{15}$ and $f(10) = \dfrac{100}{15}$, so f continuous at 10.

$\lim\limits_{x \to 5} f(x) \neq f(5)$ since $f(5)$ doesn't exist, so f is discontinuous at 5.

15. $\lim\limits_{x \to 1} f(x) = \lim\limits_{x \to 1} \dfrac{(x+1)(x-1)}{-2(x-1)} = \lim\limits_{x \to 1} \dfrac{x+1}{-2} = \dfrac{2}{-2} = -1$, so define $f(1) = -1$.

18. $f(x) = \dfrac{(x-3)(x+2)}{(x-3)(x+1)} = \dfrac{x+2}{x+1}$ $(x \neq 3)$.   f is discontinuous at 3 and -1.

$\lim\limits_{x \to 3} f(x) = \lim\limits_{x \to 3} \dfrac{x+2}{x+1} = \dfrac{5}{4}$, so $f(3) = \dfrac{5}{4}$ removes the discontinuity at 3.

$\lim\limits_{x \to -1} f(x) = \lim\limits_{x \to -1} \dfrac{x+2}{x+1}$ doesn't exist, so the discontinuity at -1 cannot
be removed.

21. $f(x) = \dfrac{3(x-2)(x+2)}{(x-2)(x+2)} = 3$ $(x \neq \pm 2)$.   f is discontinuous at -2 and 2.

$\lim\limits_{x \to \pm 2} f(x) = \lim\limits_{x \to \pm 2} 3 = 3$, so $f(-2) = 3$ and $f(2) = 3$ remove the discontinuities at -2 and 2.

24. None. No real value of x will make the denominator zero.

27. f is discontinuous at each point of $(-\infty, 0]$ since $f(x)$ is undefined
there.

30. f is continuous everywhere except possibly at x=2. $f(2) = \frac{1}{4}(2)^2 = 1$.

$\lim\limits_{x\to 2^-} f(x) = \lim\limits_{x\to 2^-} \frac{x^2}{4} = 1$, and $\lim\limits_{x\to 2^+} f(x) = \lim\limits_{x\to 2^+} (3-x) = 1$, so $\lim\limits_{x\to 2} f(x) = 1$.

Therefore, $\lim\limits_{x\to 2} f(x) = f(2)$, so f is also continuous at 2.

33. None. $f'(x) = \frac{4}{3} x^{1/3}$ exists for all real x.

36. None. $f(x) = (x^2+1)^{-1}$; $f'(x) = -2x(x^2+1)^{-2}$ exists for all real x.

39. $f(x) = (2x+1)^{1/5}$; $f'(x) = \frac{1}{5}(2x+1)^{-4/5}(2)$ exists for all x except $-\frac{1}{2}$.

42. $f(x) = |x+3|$ is not differentiable at x=-3 since there is a corner point there. See graph.

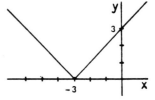

45. (a) f has a value at p, r, and t. (There is a point of the graph corresponding to each.)
    (b) The limit of f(x) exists at p, q, and t. (From each side of those points the graph approaches the same point in the plane.)
    (c) f is continuous at p and t. (For each of those, the point of the graph is the point of the plane that the graph approaches from each side.)
    (d) f is differentiable at t. (There is a nonvertical tangent there.)

48. $f(x) = |x-3|$ is continuous but not differentiable at x = 3. The graph is similar to the one with the solution of Problem 42. (3,0) is the corner point.

51. f is not continuous at x = 3 since f(3) is undefined.

54. f is continuous at x = 3.   57.

$f'(x) = \begin{cases} 1, & \text{if } x < 3 \\ 2x, & \text{if } x > 3 \end{cases}$.

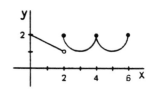

As x approaches 3 from the left,
   f'(x) is always 1.
As x approaches 3 from the right,
   f'(x) approaches 6.
Therefore, f'(3) doesn't exist.

Problem Set 3.2 Monotonicity and Concavity

3. 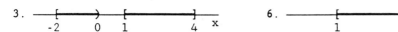   6.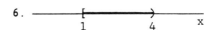

9. $2x-6 \geq 5x+1$; $-3x \geq 7$; $x \leq -\frac{7}{3}$. The solution set is $(-\infty, -\frac{7}{3}]$.

12. $15x+60 > 8x+200$; $7x > 140$; $x > 20$. The solution set is $(20, \infty)$.

15. $(x+8)(x-2) < 0$.

Values of $(x+8)(x-2)$ $\underline{\quad (+) \quad (0) \quad (-) \quad (0) \quad (+) \quad}$
                                      -8   2   x

The solution set is $(-8, 2)$. The graph is

18. $x^2-5x-6 > 0$; $(x+1)(x-6) > 0$.

Values of $(x+1)(x-6)$ $\underline{\quad (+) \quad (0) \quad (-) \quad (0) \quad (+) \quad}$
                                      -1   6   x

Sol. set is $(-\infty, -1) \cup (6, \infty)$. The graph is

21. $f'(x) = 6x-6 = 6(x-1)$. Values of $f'(x)$ $\underline{\quad (-) \quad (0) \quad (+) \quad}$
                                                          1   x

f is decreasing on $(-\infty, 1)$; increasing on $(1, \infty)$.

24. $f'(x) = -2x+8 = -2(x-4)$. Values of $f'(x)$ $\underline{\quad (+) \quad (0) \quad (-) \quad}$
                                                            4   x

f is increasing on $(-\infty, 4)$; decreasing on $(4, \infty)$.

27. $f'(x) = 3x^2-6x-12 = 3(x^2-2x-4) = 0$, if $x = 1 \pm \sqrt{5} \approx -1.2, 3.2$

Values of $f'(x)$ $\underline{\quad (+) \quad (0) \quad (-) \quad (0) \quad (+) \quad}$
                               -1.2   3.2   x

f is increasing on $(-\infty, -1.2)$ and on $(3.2, \infty)$; decreasing on $(-1.2, 3.2)$.

30. $f'(x) = 3x^2-27$; $f''(x) = 6x$. Values of $f''(x)$ _____(-)_____(0)_____(+)_____
 0 x

 f is concave down on $(-\infty,0)$; concave up on $(0,\infty)$.

33. $f'(x) = 4x^3-48x+12$; $f''(x) = 12x^2-48 = 12(x+2)(x-2) = 0$, if $x = \pm2$.

 Values of $f''(x)$ _____(+)_____(0)_____(-)_____(0)_____(+)_____
 -2 2 x

 f is concave up on $(-\infty,-2)$ and on $(2,\infty)$; concave down on $(-2,2)$.

36. $f(x) = x(x^2-27) = x(x+\sqrt{27})(x-\sqrt{27})$
 f is increasing on $(-\infty,-3)$ and on $(3,\infty)$;
 decreasing on $(-3,3)$. (Problem 26)
 f is concave down on $(-\infty,0)$;
 concave up on $(0,\infty)$. (Problem 30)

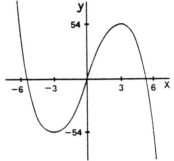

x	f(x)
±5.2	0
0	0
-3	-54
3	54

39. (a) $f'(x) = 0$ at -2, 1, and 3.7. (Tangent line is horizontal.)
 (b) $f'(x) < 0$ on $(1,3.7)$. (Graph is going downhill, left to right.)
 (c) $f''(x) > 0$ on $(-2,0)$ and on $(2,4.3)$. (Graph is concave up there.)

42. (a) $f'(x) = 0$ on $(3,4)$. (b) $f'(x) < 0$ on $(-3,0)$.
 (c) $f''(x) > 0$ nowhere.

45. 48.

51. $3x - \frac{3}{2} < 5x + \frac{10}{3}$; $18x-9 < 30x+20$; $-12x < 29$; $x > -\frac{29}{12}$.

 The solution set is $(-\frac{29}{12},\infty)$.

54. $x^2 - 7x - 18 \leq 0$; $(x+2)(x-9) \leq 0$.

Values of $(x+2)(x-9)$

(+)	(0)	(-)	(0)	(+)
	-2		9	x

The solution set is $[-2,9]$.

57. $\frac{x+3}{x-2} > 0$. Values of $\frac{x+3}{x-2}$

(+)	(0)	(-)	(u)	(+)
	-3		2	x

The solution set is $(-\infty, -3) \cup (2, \infty)$.

60. $g'(t) > 0$ and $g''(t) > 0$ from 1980 to 1990.

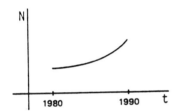

63. dy/dt increases as t increases.

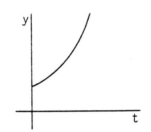

Problem Set 3.3 Extreme Points and Inflection Points

3. f has a global maximum value of $f(2) = 2$; no global minimum.

6. f has a global maximum value of $f(0) = 2$; no global minimum.

9. $f'(x) = 4x - 8 = 4(x-2) = 0$, if $x = 2$.
Critical points are 0, 3, and 2.

x	f(x)	
0	0	◁ Global maximum
3	-6	
2	-8	◁ Global minimum

12. $f'(x) = 4x - 8 = 4(x-2) = 0$, if $x = 2$.
Critical points are -1, 3, and 2.

x	f(x)	
-1	13	◁ Global maximum
3	-3	
2	-5	◁ Global minimum

Problem Set 3.3

15. $f'(x) = 6x^2 - 6x - 12 = 6(x+1)(x-2) = 0$,
   if $x = -1, 2$.

   Critical points are $-3, 3, -1,$ and 2.

x	f(x)	
-3	-43	◁ Global minimum
3	- 7	
-1	9	◁ Global maximum
2	-18	

18. $f'(x) = \frac{2}{5}x^{-3/5}$
   Critical points are $-1, 32,$ and 0.
   [0 is in domain of f but $f'(0)$ is
   undefined; $f'(x)$ is never zero.]

x	f(x)	
-1	1	
32	4	◁ Global maximum
0	0	◁ Global minimum

21. $f'(x) = (x-2)^{-2/3} - 1 = 0$, if $x = 1, 3$.

   Critical points are $-6, 3, 1,$ and 2.
   [2 is in the domain of f but $f'(2)$
   is undefined.]

x	f(x)	
-6	0	◁ Global maximum
3	0	◁ Global maximum
1	-4	◁ Global minimum
2	-2	

24. $f'(x) = \frac{-2(x+2)(x-2)}{(x^2+4)^2}$

   Critical points are $-2, 3,$ and 2.

x	f(x)	
-2	-0.5	◁ Global minimum
3	0.46	
2	0.5	◁ Global maximum

27. $f'(x) = 6x^2 - 6x = 6x(x-1)$; critical points are 0 and 1.

   Values of f'(x) (+)   (0)   (-)   (0)   (+)
   _____
                     0   1   x

   Therefore, there is a local maximum at 0; a local minimum at 1.

30. $f'(x) = \frac{2}{9}x^2 - 2 = \frac{2}{9}(x+3)(x-3)$; critical points are -3 and 3.

   Values of f'(x) (+)   (0)   (-)   (0)   (+)
   _____
                     -3   3   x

   Therefore, there is a local maximum at -3; a local minimum at 3.

33. $f'(x) = 1 - 4x^{-2} = x^{-2}(x+2)(x-2)$; critical points are -2 and 2. (0 is
   not in the domain of f.)

   Values of f'(x) (+)   (0)   (-)   (u)   (-)   (0)   (+)
   _____
                     -2   0   2   x

   Therefore, there is a local maximum at -2; a local minimum at 2.

36. $f'(x) = 3x^2+6x+3 = 3(x+1)^2$; critical point is -1.

   $f''(x) = 6x+6$; $f''(-1) = 0$, so 2nd derivative test fails. Using an auxiliary axis to look at the signs of $f'(x)$ shows that there are no local extrema since $f'(x)$ is positive on both sides of $x = -1$.

39. $f'(x) = 1+8x^{-3} = x^{-3}(x+2)(x^2-2x+4)$; critical point is -2.

   $f''(x) = -24x^{-4}$; $f''(-2) = -1.5 < 0$; therefore, local maximum at $x = -2$.

42. $f'(x) = 3x^2-27$; $f''(x) = 6x$, so 0 is a candidate.

   Values of $f''(x)$ ___(-)___(0)___(+)___
    0 x

   $f(0) = 0$. Therefore, $(0,0)$ is an inflection point.

45. $f'(x) = 4x^3-12x^2-36x$; $f''(x) = 12x^2-24x-36 = 12(x+1)(x-3)$, so -1 and 3 are candidates.

   Values of $f''(x)$ ___(+)___(0)___(-)___(0)___(+)___
    -1 3 x

   $f(-1) = -3$ & $f(3) = -179$, so $(-1,-3)$ & $(3,-179)$ are inflection points.

48. $f'(x) = 4(2x+x^{-2})$; $f''(x) = 4(2-2x^{-3}) = -8x^{-3}(x-1)(x^2+x+1)$, so 1 is a candidate. (0 is not in the domain of f.)

   Values of $f''(x)$ ___(-)___(u)___(+)___(0)___(-)___
    0 1 x

   $f(1) = 0$, so $(1,0)$ is an inflection point.

51. $f'(x) = 4x^3-12x^2 = 4x^2(x-3)$; $f''(x) = 12x^2-24x = 12x(x-2)$

   Values of $f'(x)$ ___(-)___(0)___(-)___(0)___(+)___
    0 3 x

   Values of $f''(x)$ ___(+)___(0)___(-)___(0)___(+)___
    0 2 x

x	f(x)	
0	10	◁ Inflection point (terrace point)
3	-17	◁ Global minimum
2	-6	◁ Inflection point
-1	15	
4	10	

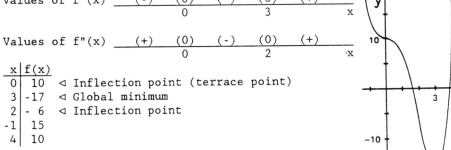

54. $f'(x) = 2x-2x^{-3/2} = 2x^{-3/2}(x^{5/2}-1)$; critical points are $\frac{1}{4}$, 4, and 1.

x	f(x)	
1/4	8.06	
4	18	◁ Global maximum
1	5	◁ Global minimum

57. $f'(x) = 3x^2-6x = 3x(x-2)$; $f''(x) = 6x-6 = 6(x-1)$

Values of $f'(x)$ (+)   (0)   (-)   (0)   (+)

              0   2   x

Values of $f''(x)$ (-)   (0)   (+)

              1   x

x	f(x)	
0	8	◁ Local maximum
2	4	◁ Local minimum
1	6	◁ Inflection point
-1	4	
-2	-12	
2	8	

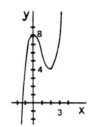

60. $f(x) = x+9x^{-1}$ is an odd function, so its graph is symmetric with respect to the origin.

$f'(x) = 1-9x^{-2} = x^{-2}(x+3)(x-3)$; $f''(x) = 18x^{-3}$

Values of $f'(x)$ (+)   (0)   (-)   (u)   (-)   (0)   (+)

             -3   0   3   x

Values of $f''(x)$ (-)   (u)   (+)

              0   x

x	f(x)	
3	6	◁ Local minimum
-3	-6	◁ Local maximum
4	6.25	
10	10.09	
1	10	
2	6.5	

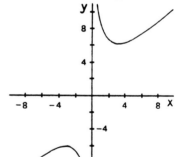

63. $f'(x)$ is 1 on (0,1), -1 on (1,2), negative on (2,3), 0 at 3, positive on (3,5), and increasing on (2,5).

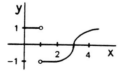

66. $f(x)$ has slope of 0 at 0, is increasing on (0,3), is constant on (3,∞), concave up on (0,1), concave down on (1,3).

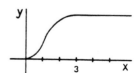

3. f(x) gets more and more positive as x gets more and more positive.

6. f(x) gets more and more positive as x gets closer and closer to 2 from the left side.

9. ∞

12. $-\infty$

15. ∞

18. $2-0 = 2$

21. $\lim\limits_{x \to \infty} \dfrac{\frac{2}{x} + \frac{5}{x^2}}{3 - \frac{2}{x^2}} = \dfrac{0 + 0}{3 - 0} = 0$

24. $\lim\limits_{x \to -\infty} \dfrac{\frac{5}{x^2} + \frac{3}{x}}{2 - \frac{4}{x^2}} = \dfrac{0 + 0}{2 - 0} = 0$

27. $\lim\limits_{x \to \infty} \dfrac{2 - \frac{1}{x} + \frac{11}{x^2}}{1 + \frac{70}{x^2}} = \dfrac{2 - 0 + 0}{1 + 0} = 2$

30. $\lim\limits_{x \to -\infty} \dfrac{1 + \frac{11}{x^2} - \frac{4}{x^3}}{5 + \frac{4}{x} + \frac{11}{x^3}} = \dfrac{1 + 0 - 0}{5 + 0 + 0} = \dfrac{1}{5}$.

33. $\lim\limits_{x \to \infty} \dfrac{5 - \frac{2}{x^{1/2}}}{1 + \frac{1}{x^{1/2}}} = \dfrac{5 - 0}{1 + 0} = 5$

36. $\lim\limits_{x \to -\infty} \dfrac{2}{\frac{4}{x^{1/3}} + 1} = \dfrac{2}{0 + 1} = 2$

39. $\lim\limits_{x \to \pm\infty} \dfrac{-3}{x^2} = 0$, so $y = 0$ is a horizontal asymptote.

   $x = 0$ is a vertical asymptote (since denominator is 0 when x is 0).

42. $\lim\limits_{x \to \pm\infty} \dfrac{2 + \frac{7}{x}}{4 - \frac{1}{x}} = \dfrac{2 + 0}{4 - 0} = \dfrac{1}{2}$, so $y = \frac{1}{2}$ is a horizontal asymptote.

   $x = \frac{1}{4}$ is a vertical asymptote (since denominator is 0 when x is $\frac{1}{4}$.)

45. $\lim\limits_{x \to \pm\infty} \dfrac{1 - \frac{4}{x^2}}{1 - \frac{2}{x} - \frac{3}{x^2}} = \dfrac{1 - 0}{1 - 0 - 0} = 1$, so $y = 1$ is a horizontal asymptote.

   $f(x) = \dfrac{(x+2)(x-2)}{(x+1)(x-3)}$. Thus, $x = -1$ and $x = 3$ are vertical asymptotes.

48. $\displaystyle\lim_{x\to\pm\infty} \dfrac{3}{2 - \dfrac{2}{x} + \dfrac{1}{x^2}} = \dfrac{3}{2 - 0 - 0} = \dfrac{3}{2}$, so $y = \frac{3}{2}$ is a horizontal asymptote.

No vertical asymptotes since $2x^2-2x+1$ is not zero for any real x.

51. Step 1: Discontinuity at x = 2.
Step 2: y = 1 is a horizontal asymptote; x = 2 is vertical asymptote.
Step 3: $f'(x) = -2(x-2)^{-2}$; $f''(x) = 4(x-2)^{-3}$

Values of f'(x) <u> (-) (u) (-) </u>   f is decreasing on $(-\infty,2)$ and
                       2   x   on $(2,\infty)$.

Values of f"(x) <u> (-) (u) (+) </u>   f is concave down on $(-\infty,2)$
                       2   x   and concave up on $(2,\infty)$.

Step 4:

x	f(x)
0	0
1	-1
-1	0.33
-2	0.5
3	3
4	2
5	1.67

Step 5:

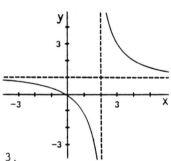

54. Step 1: Discontinuities at x = -3 and x = 3.
Step 2: y = 0 is a horizontal asymptote; x = ±3, vertical asymptotes.

Step 3: $f'(x) = \dfrac{-(x^2+2x+9)}{(x+3)^2(x-3)^2}$; $f''(x) = \dfrac{2(x^3+3x^2+27x+9)}{(x+3)^3(x-3)^3}$

Values of f'(x) <u> (-) (u) (-) (u) (-) </u>
                     -3   3   x

f is decreasing on $(-\infty,-3)$, on $(-3,3)$, and on $(3,\infty)$.

Values of f"(x) <u> (-) (u) (+) (0) (-) (u) (+) </u>
                    -3   -0.345   3   x

f is concave down on $(-\infty,-3)$ and on $(-0.345,3)$; concave up on $(-3,-0.345)$ and on $(3,\infty)$. Inflection point at x = -0.345.

Step 4:

x	f(x)
-0.345	-0.07
0	-0.11
-1	0
-2	0.2
1	-0.25
2	-0.6
-4	-0.43
4	0.71

Step 5:

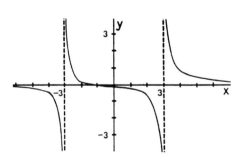

57. $f(x) = x + x^{-1}$ is an odd function so its graph is symmetric with respect to the origin.

   Step 1: Discontinuity at $x = 0$.

   Step 2: No horizontal asymptote; $x = 0$ is a vertical asymptote.

   Step 3: $f'(x) = \dfrac{(x+1)(x-1)}{x^2}$; $f''(x) = \dfrac{2}{x^3}$

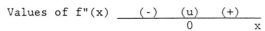

   Values of $f'(x)$ $\underline{\quad(+)\quad(0)\quad(-)\quad(u)\quad(-)\quad(0)\quad(+)\quad}$
                                -1   0   1   x

   f is inc. on $(-\infty,-1)$ and on $(1,\infty)$; dec. on $(-1,0)$ and on $(0,1)$. Local maximum at $x = -1$; local minimum at $x = 1$.

   Values of $f''(x)$ $\underline{\quad(-)\quad(u)\quad(+)\quad}$
                               0   x

   f is concave down on $(-\infty,0)$;
   concave up on $(0,\infty)$.

   Step 4:

x	f(x)
-1	-2
1	2
0.5	2.5
2	2.5
4	4.25

   Step 5:

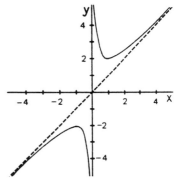

60. **Step 1:** Discontinuity at $x = 0$.

   Step 2: No horizontal asymptote; $x = 0$ is a vertical asymptote.

   Step 3: $f'(x) = \dfrac{2x^3+1}{x^2}$; $f''(x) = \dfrac{2(x-1)(x^2+x+1)}{x^3}$

   Values of $f'(x)$ $\underline{\quad(-)\quad(0)\quad(+)\quad(u)\quad(+)\quad}$
                             -0.79   0   x

   f is dec. on $(-\infty,-0.79)$; inc. on $(-0.79,0)$ and on $(0,\infty)$. Local minimum at $x = -0.79$.

   Values of $f''(x)$ $\underline{\quad(+)\quad(u)\quad(-)\quad(0)\quad(+)\quad}$
                             0   1   x

   f is concave up on $(-\infty,0)$ and on $(1,\infty)$; concave down on $(0,1)$.

   Step 4:

x	f(x)
-0.79	1.89
1	0
0.5	-1.75
2	3.5
-2	4.5
-0.5	2.25
3	8.67

   Step 5:

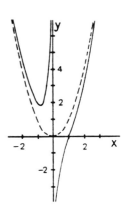

63. There are infinitely many answers. One is obtained by putting factors
    of (x+2) and (x-3) in the denominator; then let the numerator be any
    quadratic with 4 as the coefficient of x^2. For example,

    $f(x) = \dfrac{4x^2}{(x+2)(x-3)}$ is one such function.

66. **Step 1:** Discontinuities at x = -1 and at x = 4.

    Step 2: y = 2 horizontal asymptote; x = -1, x = 4 vertical asymptote.

    Step 3: $f'(x) = \dfrac{-(5x^2+14x-1)}{(x+1)^2(x-4)^2}$; $f''(x) = \dfrac{2(5x^3+21x^2-3x+31)}{(x+1)^3(x-4)^3}$

    f'(x) values (-)   (0)   (+)   (u)   (+)   (0)   (-)   (u)   (-)
                       -2.87   -1   0.07   4   x

    f is dec. on (-∞,-2.87), (0.07,4) and on (4,∞); inc. on (-2.87,-1)
    and on (-1,0.07). Local minimum at x=-2.87; local max. at x=0.07.

    Values of f"(x)   (-)   (0)   (+)   (u)   (-)   (u)   (+)
                          -4.62   -1   4   x

    f is concave down on (-∞,-4.62) and on (-1,4); concave up on
    (-4.62,-1) and on (4,∞). Inflection point at x = -4.62.

    Step 4:

x	f(x)
-2.87	1.43
0.07	0.25
-4.62	1.48
-1.4	2
-0.5	0
1	0
2	-0.83
3	-3.5
5	7.33
10	2.86

    Step 5:

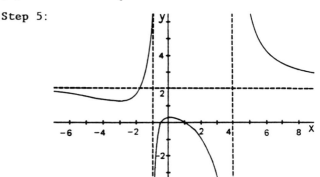

69. $\displaystyle\lim_{n\to\infty}\left[12\left(1 + \dfrac{1}{4\sqrt{n}}\right)\right] = 12(1 + 0) = 12$ minutes

72: $f(x) = \dfrac{2x^2+x+1}{x} = 2x + 1 + \dfrac{1}{x}$, so y = 2x+1 is a slant asymptote.

    Proof: $\displaystyle\lim_{x\to\pm\infty}[f(x) - (2x-1)] = \lim_{x\to\pm\infty}\dfrac{1}{x} = 0.$

    That is, for |x| large, y-values on the graph of f and
    y-values on the graph of y = 2x+1 are close.

3. (a) $-5x \le 7$; $x \ge -1.4$. The solution set is $(-1.4, \infty)$.
   (b) $2x^2 + 5x - 3 < 0$; $(2x-1)((x+3) < 0$. The solution set is $(-3, 0.5)$.

6. f is continuous and differential at $x = 1$ since f is a rational function whose denominator is not zero when x is 1.

9. f is continuous at $x = 1$ since $\lim\limits_{x \to 1} |3x-3| = 0 = f(1)$.

   $f'(1)$ would be $\lim\limits_{h \to 0} \dfrac{|3(1+h)-3| - |3-3|}{h} = \lim\limits_{h \to 0} \dfrac{3|h|}{h}$ which doesn't

   exist, so f is not differentiable at $x = 1$.

12. $\lim\limits_{x \to 1^-} f(x) = \lim\limits_{x \to 1^-} (2x^2-x+2) = 3$; $\lim\limits_{x \to 1^+} f(x) = \lim\limits_{x \to 1^+} (1.5x^2+1.5) = 3$.

   Therefore, $\lim\limits_{x \to 1} f(x) = 3$; and $f(1) = 3$, so f is continuous at $x = 1$.

   $f'(1) = \lim\limits_{h \to 0} \dfrac{f(1+h) - f(1)}{h} = \lim\limits_{h \to 0} \dfrac{f(1+h) - 3}{h} = 3$ (Consider limits from

   the left and from the right.) Thus, f is differentiable at $x = 1$.

15. $f'(x) = 3x^2 + 6x - 24 = 3(x+4)(x-2)$.

Values of f'(x)	(+)	(0)	(-)	(0)	(+)	
		-4		2		x

   Therefore, critical points are -4 (which yields a local maximum) and 2 (which yields a local minimum).

18. $f'(x) = 4x^3 - 36^2 + 96x + 4$; $f''(x) = 12x^2 - 72x + 96 = 12(x-4)(x-2)$

Values of f"(x)	(+)	(0)	(-)	(0)	(+)	
		2		4		x

   $f(2) = -80$ & $f(4) = 72$. Inflection points are $(2, -80)$ & $(4, 72)$.

21. $+\infty$ 24. $-\infty$

27. $\lim\limits_{x\to\infty} \dfrac{\dfrac{4}{\sqrt{x}} - 1}{1 + \dfrac{2}{\sqrt{x}}} = \dfrac{0-1}{1+0} = -1$

30. $f(x) = x-4x^{-2}$; $f'(x) = 1+8x^{-3} = x^{-3}(x+2)(x^2-2x+4)$; $f''(x) = -24x^{-4}$.

(a) Auxiliary axis for $f'(x)$ $\underline{\quad (+) \quad\quad (0) \quad\quad (-) \quad\quad (u) \quad\quad (+) \quad}$
$\qquad\qquad\qquad\qquad\qquad\qquad\qquad\qquad\quad -2 \qquad\qquad 0 \qquad\qquad\qquad x$

∴ f is increasing on $(-\infty,-2)$ and on $(0,\infty)$; decreasing on $(-2,0)$.

(b) Auxiliary axis for $f''(x)$ $\underline{\quad (-) \quad\quad (u) \quad\quad (-) \quad}$
$\qquad\qquad\qquad\qquad\qquad\qquad\qquad\qquad\qquad 0 \qquad\qquad x$

∴ f is concave down on $(-\infty,0)$ and on $(0,\infty)$.

(c) There is a local maximum of $f(-2) = -3$.

(d)

x	f(x)
$\sqrt[3]{4}$	0
-2	-3
2	1
4	3.75
-3	3.44
-1	-5

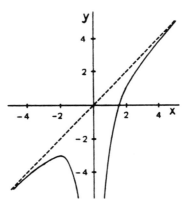

33. $f(x) = \dfrac{2x^3}{x(x+1)(x-2)}$ is one such function.

36. (a) Domain of f is $(-\infty,4) \cup (4,\infty)$, or $R \setminus \{4\}$.

(b) f is discontinuous at $x = 4$ since it is undefined there.

(c) $\lim\limits_{x\to\infty} f(x) = \lim\limits_{x\to\infty} \dfrac{\left[\dfrac{1}{x^2} - \dfrac{2}{x^3}\right]^{1/3}}{1 - \dfrac{4}{x}} = \dfrac{(0-0)^{1/3}}{1-0} = 0$

Problem Set 4.1 Optimization: Geometric Problems

3. $z = 3x-2x^2$; $z'(x) = 3-4x$.   Values of $z'(x)$ $\underline{\quad (+) \quad (0) \quad (-) \quad}$
 0.75 x

The maximum value of z occurs when x is 0.75. It is z = 1.125.

6. $z = x^2\left(\dfrac{225-3x^2}{4x}\right) = 56.25x - 0.75x^3$, $0 < x \le \sqrt{75}$; $z'(x) = -2.25(x-5)(x+5)$

Values of $z'(x)$ $\underline{\quad (+) \quad (0) \quad (-) \quad}$
 0 5 x

The maximum value of z occurs when x is 5. It is z = 187.5.

9. Let A be the area of the rectangle. $A = xy = (8-2y)y = 8y-2y^2$.

$A'(y) = 8-4y$. Values of $A'(y)$ $\underline{\quad (+) \quad (0) \quad (-) \quad}$
 0 2 4 y

The maximum area is 8. It occurs when y is 2 and x is 4.

12. $x+3y = 300$, so $x = 300-3y$. Thus, $A = xy = (300-3y)y = 300y-3y^2$.

$A'(y) = 300-6y$. Values of $A'(y)$ $\underline{\quad (+) \quad (0) \quad (-) \quad}$
 0 50 100 y

The dimensions of y = 50 meters & x = 150 meters yield a maximum area
of 7500 square meters.

15. Let C be the total cost. $xy = 1200$, so $y = 1200x^{-1}$.

$C = 6(4x+2y) + 4y = 24x+16y = 24x + 19200x^{-1}$;

$C'(x) = 24x^{-2}(x^2-800)$. Values of $A'(y)$ $\underline{\quad (-) \quad (0) \quad (+) \quad}$
 0 $20\sqrt{2}$ x

Cost will be minimum if $x = 20\sqrt{2} \approx 28.3$ ft and $y = 30\sqrt{2} \approx 42.4$ ft.

18. Let S be the surface area.

$(2x)(x)(y) = \frac{64}{3}$, so $y = \frac{32}{3}x^{-2}$.

$S = 2(2x^2+2xy+xy) = 4x^2+64x^{-1}$.

$S'(x) = 8x-64x^{-2} = 8x^{-2}(x^3-8)$.

Values of S'(x)

	(-)	(0)	(+)	
0		2		x

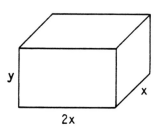

2x

S is minimum if x = 2 and $y = \frac{8}{3}$. That is, the dimensions of the chest should be 2' wide, 4' long, and 2'8" deep.

21. Let C be the total cost (relative to unit cost per foot on land).

The distance from P to B is $\sqrt{x^2+2000^2}$.

$C(x) = 2\sqrt{x^2+2000^2} + 1(10000-x)$; $C'(x) = (x^2+2000^2)^{-1/2}(2x) - 1$

$C'(x) = 0$ if $x = \frac{2000}{\sqrt{3}} \approx 1155$.

Values of C'(x)

	(-)	(0)	(+)	
0		1155		10000 x

C is minimum if x ≈ 1155. That is, B should be 1155 ft from A.

24. Let y be the length of the package,
    x be the length of an edge of a
      cross-section of the package,
and V be the volume of the package.

y+4x = 100 (largest possible).

Therefore, $V = x^2y = x^2(100-4x) = 100x^2-4x^3$.

$V'(x) = 200x-12x^2 = 4x(50-3x)$

Values of V'(x)

	(+)	(0)	(-)	
0		50/3		25 x

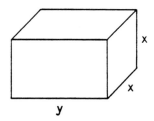

If $x = \frac{50}{3}$, then $y = \frac{100}{3}$.

The box should be $33\frac{1}{3}$" long and have a $16\frac{2}{3}$" square cross-section.

27. Let A be the sum of the areas. $0 \le x \le 10$

Radius of circle is $\dfrac{40-4x}{2\pi} = \dfrac{2(10-x)}{\pi}$.

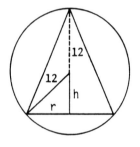

$A(x) = x^2 + \pi\left[\dfrac{2(10-x)}{\pi}\right]^2 = x^2 + \dfrac{4(x-10)^2}{\pi}$

$A'(x) = 2x + \dfrac{8(x-10)}{\pi} = \dfrac{x(2\pi+8) - 80}{\pi}$

$A'(x) = 0$ if $x = \dfrac{40}{\pi+4} \approx 5.6$, so critical points are 0, 5.6, and 10.

x	A(x)	
0	$400/\pi \approx 127$	◁ Global maximum
5.6	$400/(\pi+4) \approx 56$	◁ Global minimum
10	100	

(a) For minimum area, $x \approx 5.6$, so square uses $4(5.6) = 22.4$ cm; hence the shorter piece will be 17.6 cm (for the circle).
(b) For maximum area, the shorter piece should be 0 cm and should be used for the square (so the entire 40 cm is used for the circle).

30. Let V be the volume of the cone.

$V = \tfrac{1}{3}\pi r^2(h+12) = \tfrac{1}{3}\pi(144-h^2)(h+12)$

$\quad = \tfrac{1}{3}\pi(1728+144h-12h^2-h^3),\ 0 \le h < 12$

$V'(h) = \tfrac{1}{3}\pi(144-24h-3h^2) = -\pi(h+12)(h-4)$

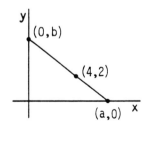

Values of $V'(h)$

	(+)	(0)	(-)	
	0	4	12	h

Radius should be $8\sqrt{2} \approx 11.3$ in; altitude 16 in.

33. Let H be the square of the length of the hypotenuse.
Note: Length of hypotenuse is minimum if H is minimum.

Using slope: $\dfrac{-b}{a} = \dfrac{2-b}{4}$, so $a = \dfrac{4b}{b-2}$.

$H = a^2+b^2 = \dfrac{16b^2}{(b-2)^2} + b^2$

$H'(b) = \dfrac{-64b}{(b-2)^3} + 2b = 0$ if $b = 2 + 2\sqrt[3]{4} \approx 5.17$

Values of $H'(b)$

	(-)	(0)	(+)	
	2	5.17		b

Length of hypotenuse is minimum when y-intercept is about 5.17 and x-intercept is about 6.52.

Problem Set 4.2   Optimization: Business Problems

3. (a) $P(x) = R(x) - C(x) = (50x-0.10x^2) - (1000+20x-0.05x^2)$
$$= -0.05x^2+30x-1000$$

   (b) $P'(x) = -0.1x+30$

   (c) $P'(x) = 0$ if $x = 300$, which yields a maximum since the graph of $P(x)$ is a parabola opening downward.

6. $p(x) = \dfrac{R(x)}{x} = 140-0.1x$.   Weekly profit was found to be maximum when

   $x = 420$ (in Problem 2). $p(420) = 140 - 0.1(420) = \$98.00$

9. Let A be the average cost.   Then $A(x) = 600x^{-1} + 8 + 0.01x^2$.

   $A'(x) = -600x^{-2} + 0.02x = x^{-2}(0.02x^3 - 600) = 0$, if $x = \sqrt[3]{30000} \approx 31.07$.

   Values of A'(x) $\dfrac{\quad (-) \quad (0) \quad (+)\quad}{0 \qquad\quad 31.07 \qquad\quad 40}$ x

   (a) $A(31) = \$36.96$; $A(32) = \$36.99$.   The minimum average cost occurs when 31 lamps are produced per day.

   (b) $\$36.96$.

12. $P(x) = R(x) - C(x) = xp(x) - C(x) = x(75-0.025x) - (3000+30x-0.015x^2)$

   $$= -0.01x^2+45x-3000$$

   $P'(x) = -0.02x+45 = 0$ if $x = 2250$, which yields a maximum since the graph of P is a parabola opening downward.   In order to maximize profit make 2250 cabinets per month.

15. (a) $R(x) = xp(x) = 1200x-x^2$

   (b) $P(x) = R(x) - C(x) = (1200x-x^2) - (30000+100x) = -x^2+1100x-30000$
   $P'(x) = -2x+1100 = 0$ if $x = 550$, which yields a maximum since the graph of P is a parabola opening downward.   Thus, to maximize profit produce 550 units per year.

   (c) $P(550) = -(550)^2 + 1100(550) - 30000 = \$272,500$

   (d) $p(550) = 1200 - (550) = \$650$

   (e) $R'(x) = 1200-2x$; $R'(200) = 1200 - 2(200) = \$800$ per unit

   (f) $\$800$ [answer for (e)]

18. Let I be inventory cost and x be the lot size.

I(x) = (average # balls stored)(storage cost per ball)
                    + [250 + (2)(#balls ordered)](#orders per year)

$$= \left(\frac{x}{2}\right)(5) + [250 + (2)(x)]\left(\frac{24000}{x}\right) = 2.5x + 6000000x^{-1} + 48000$$

$I'(x) = 2.5 - 6000000x^{-2} = 0$, if $x = \sqrt{2400000} \approx 1549$

Values of I'(x)
$$\frac{\quad (-) \quad (0) \quad (+) \quad}{0 \qquad 1549 \qquad x}$$

Lots of size 1549 will minimize inventory cost.

21. Let P be profit made on customers' savings.
$P(x) = R(x) - D(x) = 0.14(kx^{3/2}) - (kx^{3/2})x = k(0.14x^{3/2} - x^{5/2})$
$P'(x) = k(0.21x^{1/2} - 2.5x^{3/2}) = kx^{1/2}(0.21 - 2.5x) = 0$, if $x = 0.084$

Values of P'(x)
$$\frac{\quad (+) \quad (0) \quad (-) \quad}{0 \qquad 0.084 \qquad 0.14 \;\; x}$$ Offer 8.4% interest.

24. Let P be the profit North Star makes on customers' deposits, and let x be the interest rate they offer to customers.
$P(x) = [k(x-0.04)^2](0.10) - [k(x-0.04)^2]x = k(x-0.04)^2(0.10-x)$
$P'(x) = -3k(x-0.04)(x-0.08)$

Values of P'(x)
$$\frac{\quad (+) \quad (0) \quad (-) \quad}{0.04 \quad 0.08 \qquad 0.10 \;\; x}$$ Offer 8% interest.

27. Let x be the number of weeks Susan waits to market her steer, and let A be the amount (in cents) she will get for the steer.

$A(x) = (600+30x)(60-2x) = 60(600+10x-x^2)$

$A'(x) = 60(10-2x) = 0$ if $x = 5$, which yields a maximum since the graph of A is a parabola opening downward. In order to maximize the amount she will get, she should wait 5 weeks.

30. Let x be the number of days John waits, and R be revenue (in cents).

$R(x) = (1000+30x)(250-5x) = 50(5000+50x-3x^2)$

$R'(x) = 50(50-6x) = 0$ if $x \approx 8.3$ days, which yields a maximum since the graph of R is a parabola opening downward. Therefore, since $R(8) = 260400$ cents and $R(9) = 260350$ cents, John should harvest and sell his crop in 8 days.

Problem Set 4.3 More Applications

3. Let t be the number of additional trees on an acre and y be the yield per acre. Then $y(t) = (40+t)(60-0.75t) = 2400+30t-0.75t^2$, $t \in [0,15]$.

   (a) $y(12) = 2400 + 360 - 108 = 2652$ lbs.

   (b) $y'(t) = 30-1.5t = 0$ if $t = 20$.

   Values of $y'(t)$

   ```
                        (+)
   _____
   0                          15   t
   ```

   y is increasing on $[0,15]$ so plant 15 more trees (total of 55 trees).

6. Let x be the number of 10-passenger groups in excess of 300 persons and let R be the total revenue from a group consisting of at least 300 persons but also being a multiple of 10.

   $R(x) = [10(30+x)][15-0.2x] = 4500+90x-2x^2$, for $x = 0,1,2,\cdots,75$

   $R'(x) = 90-4x = 0$, if $x = 22.5$, and since the graph of R is parabolic and opening downward, R is maximum at $x = 22.5$ (but x must be a whole number).

   $R(22) = \$5,512$ and $R(23) = \$5,512$; thus, if the total group size is a multiple of 10, then the maximum revenue is \$5,512 and occurs for 300 plus either 22 or 23 groups of 10 more passengers, i.e., for either 520 or 530 passengers. Thus, the maximum revenue for groups of **any** size will occur for 300 plus 22 groups of 10 plus 9 passengers, for a total of 529 passengers and a total revenue of $529[15 - 0.2(22)]$ which is $529(\$10.60) = \$5,607.40$.

9. Let C be the cost of fuel for the trip.

   $C(x) = (3x^{-1} + 0.0015x)(400)(1.30) = 1560x^{-1} + 0.78x$

   $C'(x) = -1560x^{-2} + 0.78 = 0$ if $x = \sqrt{2000} \approx 45$ mph

   Values of $C'(x)$

   ```
               (-)     (0)     (+)
   _____
   0                   45            x
   ```

   (a) Drive at about 45 mph.   (b) $C(\sqrt{2000}) = \$69.77$

12. $R'(x) = \dfrac{2Cx-3x^2}{a} = 0$, if $x = 0, \dfrac{2C}{3}$; $R''(x) = \dfrac{2C-6x}{a} = 0$, if $x = \dfrac{C}{3}$.

   Values of $R'(x)$

   ```
            (+)     (0)     (-)                R''(x)    (+)     (0)     (-)
   _____          _____
   0        2C              C    x       0                C              C    x
            3                                              3
   ```

   R' is the sensitivity. It is maximum for a dosage of C/3, which is one-half of 2C/3, the dosage for which the reaction is maximum.

15. $S'(t) = 60t-9t^2 = 0$, if $t = 0$, $\frac{20}{3}$; $S''(t) = 60-18t = 0$, if $t = \frac{10}{3}$.

Values of $S'(t)$ $\underbrace{\quad (+) \quad}_{0} \underbrace{(0)}_{\frac{20}{3}} \underbrace{(-) \quad}_{7}$ t $S''(t)$ $\underbrace{\quad (+) \quad}_{0} \underbrace{(0)}_{\frac{10}{3}} \underbrace{(-) \quad}_{7}$ t

(a) Score is maximum for about 6 hours, 40 minutes of study time.

(b) Rate of learning, $S'(t)$, is maximum after 3 hours, 20 minutes.

18. Let x be the number of 5-cent increases & R be the daily revenue ($).

$R(x) = (80000 - 1000x)(1.50 + 0.05x) = 120000 + 2500x - 50x^2$

$R'(x) = 2500-100x = 0$ if $x = 25$, which yields a maximum since the graph of R is a parabola opening downward. In order to maximize daily revenue, increase the toll by 5(25) = 125 cents to $2.75.

21. $C(t) = 600[1 - 8(t+4)^{-1} + 64(t+4)^{-2}]$

$C'(t) = 600[8(t+4)^{-2} -128(t+4)^{-3}] = 4800(t+4)^{-3}(t-12)$

$C''(t) = [4800(-3)(t+4)^{-4}](t-12) + [4800(t+4)^{-3}](1) = -9600(t+4)^{-4}(t-20)$

Values of $C'(t)$ $\underbrace{\quad (-) \quad}_{0} \underbrace{(0)}_{12} \underbrace{(+) \quad}_{t}$ $C''(t)$ $\underbrace{\quad (+) \quad}_{0} \underbrace{(0)}_{20} \underbrace{(-) \quad}_{t}$

(a) Oxygen concentration will be minimum after 12 days.

(b) Rate of change of oxygen concentration, $C'(t)$, will be maximum at the end of 20 days.

24. Let x be the increase in the ticket price ($) and R be total revenue.

$R(x) = (60000-6000x)(x+7.25) = -6000x^2 + 16500x + 435000$

$R'(x) = -12000x + 16500 = 0$ if $x = 1.375$, which yields a maximum since the graph of A is a parabola opening downward. Thus, to maximize the total revenue increase the ticket price by $6.37 or $6.38. Each will yield a total revenue of $446,634.60. However, it is likely that the increase would be a multiple of 25 cents, in which case the optimum price would be $6.25 or $6.50, each yielding $446,250.00. If multiples of a dollar are desired, the optimum price would be $6.00, yielding $445,500. [A $7.00 price would yield only $444,000.]

3. $y(4+3x) = 2-x$

   $y = \dfrac{2-x}{4+3x}$

6. $x^2y-2y = 5x-9$

   $y(x^2-2) = 5x-9$

   $y = \dfrac{5x-9}{x^2-2}$

9. $y^2 = x^2-2x+4$

   $y = \pm\sqrt{x^2-2x+4}$

12. $y^2+2y \quad = 2x-4$

   $y^2+2y+1 = 2x-3$

   $(y+1)^2 = 2x-3$

   $y+1 = \pm\sqrt{2x-3}$

   $y = -1 \pm \sqrt{2x-3}$

15. $(2x^2)(y) = 7$

   $(2x^2)(y') + (y)(4x) = 0$

   $y' = \dfrac{-4xy}{2x^2} = \dfrac{-2y}{x}$

18. $(3x)(y') + (y)(3) - 2 = 2x + 2y'$

   $(3x-2)y' = 2x-3y+2$

   $y' = \dfrac{2x-3y+2}{3x-2}$

21. $2x + (2xy' + 2y) + 6yy' = 0$

   $(2x+6y)y' = -2x-2y$

   $y' = \dfrac{-2x-2y}{2x+6y} = \dfrac{-(x+y)}{x+3y}$

24. $5(3+y^2)^4(2yy') = 200x$

   $y' = \dfrac{200x}{10y(3+y^2)^4} = \dfrac{20x}{y(3+y^2)^4}$

27. **Point:** It is given to be $(2,1)$.

   Slope: $x^2+4y^2 = 4y+2x$

   $2x + 8yy' = 4y'+2$

   $(8y-4)y' = 2-2x$

   $y' = \dfrac{2-2x}{8y-4} = \dfrac{1-x}{4y-2}$; $\quad y(2,1) = \dfrac{1-2}{4-2} = -\dfrac{1}{2}$ is the slope at $(2,1)$.

   Equation: $y-1 = -\dfrac{1}{2}(x-2)$ or $y = -\dfrac{1}{2}x + 2$ or $x+2y = 4$

30. **Point:** It is given to be $(-2,1)$.

   Slope: $y^3-2xy = 5$

   $3y^2y' - (2xy' + 2y) = 0$

   $(3y^2-2x)y' = 2y$

   $y' = \dfrac{2y}{3y^2-2x}$; $\quad y(-2,1) = \dfrac{2}{3+4} = \dfrac{2}{7}$ is the slope at $(-2,1)$.

   Equation: $y-1 = \dfrac{2}{7}(x+2)$ or $y = \dfrac{2}{7}x + \dfrac{11}{7}$ or $2x-7y = -11$

33. $x^2+y^2 = 50^2$ and $\frac{dy}{dx} = -2$ ft/sec

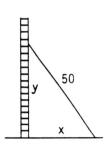

$2x \frac{dx}{dt} + 2y \frac{dy}{dt} = 0$; $2x \frac{dx}{dt} + 2y(-2) = 0$,

so $\frac{dx}{dt} = \frac{2y}{x}$.

If $y = 30$, then $x = 40$ and so $\frac{dx}{dt} = \frac{2(30)}{(40)} = 1.5$ ft/sec.

36. $s^2 = x^2+(295)^2$ and $\frac{dx}{dt} = 30$ ft/sec

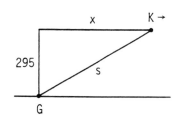

$2s \frac{ds}{dt} = 2x \frac{dx}{dt} + 0$; $2s \frac{ds}{dt} = 2x(30) = 60x$,

so $\frac{ds}{dt} = \frac{30x}{s}$.

If $s = 500$, then $x = \sqrt{162975}$ and so $\frac{ds}{dt} = \frac{30\sqrt{162975}}{500} \approx 24.2$ ft/sec.

39. (a) We need to find $\frac{dx}{dt}$ when $p = 6$ and $\frac{dp}{dt} = -0.08$.

$px + 12p = 600$, so $p \frac{dx}{dt} + x \frac{dp}{dt} + 12 \frac{dp}{dt} = 0$

If $p = 6$, then $x = 88$ and so

$(6)\frac{dx}{dt} + (88)(-0.08) + 12(-0.08) = 0$; $\frac{dx}{dt} = \frac{4}{3}$ lb/wk.

(b) Let R be the revenue. $R = px$, so $\frac{dR}{dt} = p \frac{dx}{dt} + x \frac{dp}{dt}$.

Therefore, at the given time, $\frac{dR}{dt} = (6)(\frac{4}{3}) + (88)(-0.08) = 0.96$, so

revenue is increasing at about 96 cents per week.

42. $y = 15 + 0.01x^2$, $x \in [-20,20]$, and $\frac{dx}{dt} = 4$ ft/sec are given.

$\frac{dy}{dt} = 0.02x \frac{dx}{dt} = 0.08x$; if $x = 10$, $\frac{dy}{dt} = 0.08(10) = 0.8$ ft/sec.

45. Let V be the volume of the tumor. $V = \frac{4}{3}\pi r^3$ and $\frac{dr}{dt} = 0.2$ cm/day.

$$\frac{dV}{dt} = 4\pi r^2 \frac{dr}{dt} = 4\pi r^2(0.2) = 0.8\pi r^2. \quad \text{If } r = 8, \frac{dV}{dt} = 0.8\pi(8)^2 \approx 161.$$

The tumor is growing today at about 161 cubic centimeters per day.

48. $x = 4500 - 1.2p$ and $\frac{dp}{dt} = -150$, so $\frac{dx}{dt} = -1.2\frac{dp}{dt} = -1.2(-150) = 180$.

Let R be the revenue. $R = xp$, so $\frac{dR}{dt} = x\frac{dp}{dt} + p\frac{dx}{dt} = x(-150) + p(180)$.

If $p = 1800$, $x = 2340$, so $\frac{dR}{dt} = (2340)(-150) + (1800)(180) = -27000$.

Revenue is decreasing at \$27,000 per year.

Problem Set 4.5   Approximations and the Differential

3. $dy = (2x^{-1/2} - 5x^{-2})dx$

6. $y = (x^2+3x2+2)^{1/2}$;
$dy = \frac{1}{2}(x^2+3x+2)^{-1/2}(2x+3)dx$

9. $y = \sqrt{x}$, so $dy = \dfrac{1}{2\sqrt{x}} dx$.

For $x_1 = 4$ and $x_2 = 4.02$: $dx = 4.02 - 4 = 0.02$
$\Delta y = \sqrt{4.02} - \sqrt{4} \approx 0.00499$
$dy = \dfrac{1}{2\sqrt{4}} = 0.005$

12. $dy = 3(3 + 2\sqrt{x})^2[2(1/2\sqrt{x})]dx$

For $x_1 = 1$ and $x_2 = 0.98$: $dx = 0.98 - 1 = -0.02$
$\Delta y = (3 + 2\sqrt{0.98})^3 - (3 + 2\sqrt{1})^3 \approx -1.502$
$dy = 3(3 + 2\sqrt{1})^2[2(1/2\sqrt{1})](-0.02) = -1.5$

15. Let $y = x^{1/3}$; then $dy = \frac{1}{3}x^{-2/3}dx$.

For $x_1 = 27$ and $x_2 = 26.75$: $dx = 26.75 - 27 = -0.25$
$dy = \frac{1}{3}(27)^{-2/3}(-0.25) \approx -0.0093$
Therefore, $N = (26.75)^{1/3} \approx (27)^{1/3} + (-0.0093) = 2.9907$

18. Let $y = [2\sqrt{x} + 3x^2]^3$; then $dy = 3[2\sqrt{x} + 3x^2]^2(1/\sqrt{x} + 6x)dx$.

For $x_1 = 1$ and $x_2 = 1.01$: $\quad dx = 1.01 - 1 = 0.01$
$$dy = 3[2\sqrt{1} + 3(1)^2]^2[1/\sqrt{1} + 6(1)](0.01) = 5.25$$

Therefore, $N = [2\sqrt{1.01} + 3(1.01)^2]^3 \approx [2\sqrt{1} + 3(1)^2]^3 + (5.25) = 130.25$

21. $c^2 = x^2 + (2x)^2 = 5x^2$, so $c = \sqrt{5}x$ and $dc = \sqrt{5}dx$.

For $x = 48.62$, let $dx = 0.21$ (absolute value maximum error).
Then $dc = \sqrt{5}(0.21) \approx 0.47$, so $c = \sqrt{5}(48.62) \pm 0.47 \approx 108.72 \pm 0.47$.

24. Answers for Problem 22 were $V = 14887 \pm 726$ and $s = 3631 \pm 118$. Thus, relative errors are $\dfrac{726}{14887} \approx 0.0488 = 4.88\%$ for the volume, and $\dfrac{118}{3631} \approx 0.0325 = 3.25\%$ for the surface area.

27. $x = f(p) = 500 - 1.5p$; $f'(p) = -1.5$

(a) $E(p) = -\dfrac{pf'(p)}{f(p)} = -\dfrac{p(-1.5)}{500-1.5p} = \dfrac{1.5p}{500-1.5p}$, so $E(12) = \dfrac{18}{482} < 1$.

Therefore, the demand is inelastic.

(b) Price should be increased to increase revenue.

(c) $dR = f(p)[1 - E(p)]dp$

For $p = 1$ and $dp = 0.10$, $dR = f(1)[1 - E(12)](0.10)$
$$= (482)[1 - \tfrac{18}{488}](0.10) = 46.42.$$
Revenue will increase by about \$46.40 per week.

30. $x = f(p) = 5625 - 3p^2$; $f(p) = -6p$

(a) $E(p) = -\dfrac{pf'(p)}{f(p)} = -\dfrac{p(-6p)}{5625-3p^2} = \dfrac{2p^2}{1875-p^2}$

(b) $\dfrac{2p^2}{1875-p^2} > 1$ if $p > 25$ and $p < \sqrt{1875}$

Demand is elastic if $25 < p < \sqrt{1875} \approx 43.3$.

(c) Relative change in demand is $\dfrac{dx}{x} = \dfrac{f'(p)dp}{f(p)} = \dfrac{-6pdp}{5625-3p^2}$ which equals

-0.0113 or -1.13% when $p = \$10$ and $dp = \$1$ (10% of \$10).

33. Let $y = f(x) = x^{1/3} + 5x^2 - 280$, so $dy = (\frac{1}{3}x^{-2/3} + 10x)dx$

For $x = 8$ and $dx = 0.01$, $dy = [\frac{1}{3}(8)^{-2/3} + 10(8)](0.01) \approx 0.801$.

Therefore, $f(8.01) \approx f(8) + dy = [\sqrt[3]{8} + 5(8)^2 - 280] + 0.801 = 42.801$.

36. (a) Let P be profit.
$P = xp - C(x) = x(1200-x) - (3000+400x) = -x^2+800x-3000$
$dP = (-2x+800)dx$ which equals 0 when $x = 400$ and $dx = 10$.
That is, profit will increase by about \$4,000.

(b) Let the average cost be $A(x) = \dfrac{C(x)}{x} = 3000x^{-1} + 400$

$dA = -3000x^{-2}dx$ which equals -0.1875 when $x = 400$ and $dx = 10$.
That is, average cost will decrease by about 18.75 cents.

(c) $x = f(p) = 1200-p$, so $f'(p) = -1$.
$E(p) = \dfrac{-pf'(p)}{f(p)} = \dfrac{-p(-1)}{1200-p} = \dfrac{p}{1200-p} > 1$ if $p > 600$ and $p < 1200$.
That is, demand is elastic if $600 < p < 1200$.

Chapter 4 Review Problem Set

3. Let z be the square of the distance between $(10,0)$ and $(x,\sqrt{x})$.

$z(x) = (x-10)^2+(\sqrt{x}-0)^2 = x^2-19x+100$, $x \geq 0$

$z'(x) = 2x-19$ which is 0 if $x = 9.5$
$z''(x) = 2 > 0$, so $x = 9.5$ yields a minimum.

The points of the parabola closest to $(10,0)$ are $(9.5,\sqrt{9.5})$ and, by symmetry, $(9.5,-\sqrt{9.5})$.

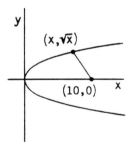

6. Let A be the area of the window. Since the perimeter is 20 we have
$20 = x+2y+\pi(\frac{1}{2}x)$, so $y = 10-\frac{1}{2}x-\frac{1}{4}\pi x$.

Thus, $A = xy + \frac{1}{2}\pi(\frac{1}{2}x)^2 = x(10-\frac{1}{2}x-\frac{1}{4}\pi x) + \frac{1}{2}\pi(\frac{1}{2}x)^2 = 10x-\frac{1}{2}x^2-\frac{1}{8}\pi x^2$.

We want to maximize $L(x) = x(10-\frac{1}{2}x-\frac{1}{4}\pi x) + \frac{1}{2}[\frac{1}{2}\pi(\frac{1}{2}x)^2] = 10x-\frac{1}{2}x^2-\frac{3}{16}\pi x^2$.

$L'(x) = 10-x-\frac{3}{8}\pi x = 0$ if $x = \frac{80}{8+3\pi} \approx 4.59$, which yields a maximum since the graph of L is a parabola opening downward. That is, the maximum amount of light is allowed if $x \approx 4.59$ ft. The corresponding value of y is about 4.10 ft.

9. (a) $A(x) = \dfrac{C(x)}{x} = 1000x^{-1}+50+0.07x$, $0 \le x \le 200$, is the average cost.

$A'(x) = -1000x^{-2}+0.07$ is 0 if $x = \sqrt{1000/0.07} \approx 119.52$

Values of $A'(x)$

	(-)	(0)	(+)	
0		119.52		x

(b) $A(119) = 66.73$ and $A(120) = 66.73$, so produce 119 or 120 sets per week for a minimum average cost of $66.73.

12. (a) Profit is $P(x) = 105x - (1200+60x+0.08x^2) = -0.08x^2+45x-1200$
$P'(x) = -0.16x+45$ is 0 if $x = 281.25$
$P''(x) = -0.16 < 0$, so $x = 281.25$ yields a maximum.

(b) $P(281) = 5128.12$ and $P(282) = 5128.08$, so a maximum profit of $5,128.12 is realized if 281 houses are produced weekly.

15. Let I = (holding cost) + (reordering cost) be the inventory cost.
I = (average number held for a year)($2)
+ (number of reorders in a year)[$100 + (reorder size)($1)]
Assume the avg. no. held for a year is $\frac{x}{2}$.

Then, $I(x) = (\frac{x}{2})(2) + (\frac{6000}{x})[100+(x)(1)] = x + 600000x^{-1} + 6000$

$I'(x) = 1 - 600000x^{-2}$ is 0 if $x = \sqrt{600000} \approx 774.6$

Values of $I'(x)$

	(-)	(0)	(+)	
0		774.6		x

$I(774) = 7549.19$, and $I(775) = 7549.19$. Note that $I(800) = 7550$ and $I(768) = 7549.25$ (768 is 64 dozen), so the inventory cost is not changing very much near the minimum. Therefore, order in convenient lot sizes somewhere in the neighborhood of 775 per lot.

18. Let R be revenue from dues and x be the amount of decrease in dues.
$R(x) = (120+2x)(36-x)$, $x \ge 0$; $R'(x) = -4x-48$ is 0 if $x = -12$.

Since the graph of R is a parabola opening downward with vertex where x is -12, R is decreasing for $x \ge 0$. Therefore, maximum revenue is achieved by leaving the dues right where they are. [Note that if we were allowed to increase dues and the result were the *loss* of two members for each $1 increase, then revenue would be maximum if dues were increased by $12, resulting in a club of 96 members.]

21. $[(1)(y^3) + (x)(3y^2y')] + [(2x)(y) + (x^2)(y')] = 3$

$y^3+3xy^2y'+2xy+x^2y' = 3$

At $(1,2)$: $8+12y'+4+y' = 3$; $y' = -\frac{9}{13}$ is the slope of the tangent.

Equation: $y-2 = -\frac{9}{13}(x-1)$ or $9x+13y = 35$ or $y = -\frac{9}{13}x+\frac{35}{13}$

24. Let t be the time after 12:30 p.m.
Let z be the distance between the planes.

$z^2 = [400(t+\frac{1}{2})]^2 + [560t]^2$

$2z\frac{dz}{dt} = 800(t+\frac{1}{2})(400) + 2[560t](560)$

At 1:30 p.m., $t = 1$, $z = \sqrt{(600)^2+(560)^2} = \sqrt{673700}$

$\therefore 2\sqrt{673700}\frac{dz}{dt} = 800(\frac{3}{2})(400) + 2[560](560)$, so $\frac{dz}{dt} \approx 674.52$ mph

27. (a) $p = 8 - 0.005x$

Solving for x, $x = f(p) = 200(8-p)$; then $f'(p) = -200$

$E(p) = -\frac{pf'(p)}{f(p)} = -\frac{p(-200)}{200(8-p)} = \frac{p}{8-p}$

Therefore, $E(6.5) = \frac{6.5}{1.5} > 1$, so the demand is elastic.

(b) Decrease the price.

(c) $dR = f(p)[1-E(p)]dp$

For $p = 6.5$ and $dp = 0.5$, $dR = f(6.5)[1-E(6.5)](0.5)$

$= 200(8-6.5)[1-\frac{13}{3}](0.5) = -500$,

so the weekly revenue will decrease by $500.

30. (a) $f'(x) = \frac{(x)[2(x+2)] - (x+2)^2(1)}{x^2} = \frac{x^2-4}{x^2} = 1-4x^{-2}$

$f'(\frac{1}{2}) = -15$ is the slope of the tangent where x is $\frac{1}{2}$.

(b) $f''(x) = 8x^{-3} > 0$ if $x > 0$, so f is concave up on its domain.

30. (continued)

(c) $\lim\limits_{x\to\infty}$ [f(x) - (x+4)] = $\lim\limits_{x\to\infty}\left[\dfrac{4}{x}\right]$ = 0, so y = x+4 is an asymptote.

(d) Values of f'(x)

Therefore, the global minimum value of f is f(2) = 8.

(e)

x	f(x)
1	9
2	8
4	9
8	12.5

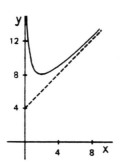

33. $\lim\limits_{x\to\sqrt{2}}\dfrac{(2x-3)(x-\sqrt{2})}{x^2-2}$ = $\lim\limits_{x\to\sqrt{2}}\dfrac{(2x-3)(x-\sqrt{2})}{(x+\sqrt{2})(x-\sqrt{2})}$ = $\lim\limits_{x\to\sqrt{2}}\dfrac{2x-3}{x+\sqrt{2}}$ = $\dfrac{2\sqrt{2}-3}{2\sqrt{2}}$ ≈ -0.0607

36. f(x) = $[2(x+7)^{1/2} - 4(x-1)^{-2}]^4$

f'(x) = $4[2(x+7)^{1/2} - 4(x-1)^{-2}]^3[(x+7)^{-1/2} + 8(x-1)^{-3}]$

f'(2) = $4[2(9)^{1/2} - 4(1)^{-2}]^3[(9)^{-1/2}+8(1)^{-3}]$ = $\dfrac{800}{3}$

Problem Set 5.1   Exponential and Logarithmic Functions

3. $5^1 = 5$

6. $4^{-3}4^4 = 4^1 = 4$

9. 0.0354 (approximately)

15. $b^{-6}b^7 = b^1 = b$

18. $c^{-2}c^6c^{-3} = c^1 = c$

21. $(\pi^4)^{-2} = \pi^{-8} = 1/\pi^8$

24. $(b-1)^{11/12}$

27. 5

30. $\frac{3}{2}$

12.

33. -0.3644 (approx.)   36. 0.8879 (approx.)

39. $log_a(2^23) = 2log_a2 + log_a3 \approx 2(0.693) + 1.099 = 2.485$

42. $log_a5 + log_a2^{1/2} = log_a5 + \frac{1}{2}log_a2 \approx 1.609 + \frac{1}{2}(0.693) \approx 1.956$

45. $\dfrac{log_a5 + log_a(a^2)}{log_a(3^4)} = \dfrac{log_a5 + 2log_aa}{4log_a3} \approx \dfrac{1.609 + 2(1)}{4(1.099)} \approx 0.821$

48. $log_b(\frac{xy}{z})$

51. $log_b \dfrac{x^3z^6}{y^3}$

54.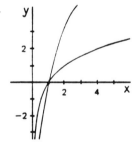

57. $2^x = 16$; $x = 4$
    The only solution is 4.

60. $x = \pi$
    The only solution is π.

63. $5^2 = x+2$; $x+2 = 25$; $x = 23$
    The only solution is 23.

66. $2^{3x+2} = 2^6$; $3x+2 = 6$
    The only solution is $\frac{4}{3}$.

69. $log_{10}(10^{10000}) = 10000$
    $log_{10}(13^{9000}) = 9000log_{10}13 \approx 10025$
    Thus, $13^{9000} > 10^{10000}$

72. $1 + 2 + 2^2 + \cdots + 2^{63} = 2^{64}-1$ grains $\approx 369,000,000,000,000$ kg. [It
is interesting to note that the world's rice production for 1986 was
only about $474,000,000,000$ kg (*Statistical Abstract of the United
States 1989*, page 811); at that production level it would have taken
the prince about 778 years to satisfy the request.]

3. -4.6460

6. 6.8533

9. 24.0468

12. 22.4592

15. $\ln(3^x) = \ln 15$
    $x\ln 3 = \ln 15$
    $x = \dfrac{\ln 15}{\ln 3} \approx 2.4650$

18. $\ln(4^{x-3}) = \ln(9.82)$
    $(x-3)\ln 4 = \ln(9.82)$
    $(x-3) = \dfrac{\ln(9.82)}{\ln 4} \approx 1.6479$
    $x \approx 4.6479$

21. $\sqrt{x}\,\ln 4 = \ln 10$
    $\sqrt{x} = \dfrac{\ln 10}{\ln 4}$
    $x \approx 2.7588$

24. $(x^2+x)\ln 3 = \ln 7$
    $x^2 + x - \dfrac{\ln 7}{\ln 3} = 0$
    $x = \dfrac{-1 \pm \sqrt{1 + (4\ln 7)/(\ln 3)}}{2}$
    $x \approx -1.9217$ or $x \approx 0.9217$

27. **Point:**   If $x = e$, $y = 3$.

    Slope: $\dfrac{dy}{dx} = \dfrac{3}{x}$ which equals $\dfrac{3}{e}$ when $x = e$, so the slope is $\dfrac{3}{e}$

    Equation: $y-3 = \dfrac{3}{e}(x-e)$ or $y = \dfrac{3}{e}x$ [Approximately, $y = 1.1036x$]

30. **Point:**   If $x = 1$, $y = 1$.

    Slope: $\dfrac{dy}{dx} = 3(x^2 + \ln x)^2(2x + \dfrac{1}{x})$ which equals 9 (slope) when $x = 1$.

    Tangent: $y-1 = 9(x-1)$ or $y = 9x-8$

33. $f'(x) = (x^2)(\dfrac{1}{x}) + (\ln x)(2x)$

    $= 2x\ln x + x = x(1 + 2\ln x)$

36. $f'(x) = \dfrac{1}{2}(2 + \ln x)^{-1/2}\dfrac{1}{x}$

39. $f'(x) = 3(x^2 + 2\ln x)^2(2x + \dfrac{2}{x})$

42. $f'(x) = (3x^2)(e^x) + (e^x)(6x)$

    $= 3xe^x(x+2)$

45. $f'(x) = \dfrac{1}{5}(5e^x + x^5)^{-4/5}(5e^x + 5x^4) = (5e^x + x^5)^{-4/5}(e^x + x^4)$

48. (a) $f'(x) = \dfrac{2xe^x - x^2e^x}{(e^x)^2} = \dfrac{2x-x^2}{e^x} = \dfrac{-x(x-2)}{e^x}$

Values of $f'(x)$

	(-)	(0)	(+)	(0)	(-)	
		0		2		x

f is decreasing on $(-\infty,0)$ and on $(2,\infty)$; increasing on $(0,2)$.

(b) f has no maximum value since $\displaystyle\lim_{x\to-\infty} f(x) = \infty$.

(c) $f''(x) = \dfrac{(e^x)(2-2x) - (2x-x^2)(e^x)}{(e^x)^2} = \dfrac{x^2-4x+2}{e^x} = 0$ if $x \approx 0.58,\ 3.41$

Values of $f''(x)$

	(+)	(0)	(-)	(0)	(+)	
		0.58		3.41		x

f is concave up on $(-\infty,0.58)$ and on $(3.41,\infty)$; down on $(0.58,3.41)$.

(d)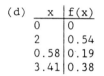

x	f(x)
0	0
2	0.54
0.58	0.19
3.41	0.38

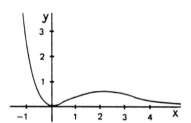

51. $\ln(3x)^2 + \ln(x^2-1) - \ln(x-1) - \ln x$

$= \ln\left[\dfrac{9x^2(x^2-1)}{x(x-1)}\right] = \ln\left[\dfrac{9x^2(x+1)(x-1)}{x(x-1)}\right] = \ln[9x(x+1)] \quad (x > 1)$

54. (a) $g'(x) = (x^3)(e^x) + (e^x)(3x^2) = x^2e^x(x+3)$

$g''(x) = (2x)[e^x(x+3)] + (e^x)[(x^2)(x+3)] + (1)[x^2(x+3)]$
$= xe^x(x^2+6x+6)$

(b) $g'(x) = (x)[3(x+\ln x)^2(1+\frac{1}{x})] + [(x\ln x)^3](1)$
$= (x+\ln x)^2(4x+3+\ln x)$

$g''(x) = (x+\ln x)^2(4+\frac{1}{x}) + (4x+3+\ln x)[2(x+\ln x)(1+\frac{1}{x})]$

$= \frac{1}{x}(x+\ln x)(12x^2+15x+6+6x\ln x+3\ln x)$

Problem Set 5.3   The Chain Rule

3. $\dfrac{dy}{dx} = e^{x^2}(2x) = 2xe^{x^2}$

6. $\dfrac{dy}{dx} = 3\,\dfrac{1}{1-x^2}\,(-2x) = \dfrac{6x}{x^2-1}$

9. $\dfrac{dy}{dx} = 3e^{4\sqrt{x}}\,\dfrac{4}{2\sqrt{x}} = \dfrac{6e^{4\sqrt{x}}}{\sqrt{x}}$

12. $\dfrac{dy}{dx} = \dfrac{2x-1}{x^2}\,\dfrac{(2x-1)(2x) - (x^2)(2)}{(2x-1)^2}$

$\qquad = \dfrac{2(x-1)}{x(2x-1)}$

15. $\dfrac{dy}{dx} = \dfrac{1}{(2x^2+5)(3x-2)}\,[(2x^2+5)(3) + (3x-2)(4x)] = \dfrac{18x^2-8x+15}{(2x^2+5)(3x-2)}$

18. $f'(x) = (2x+1)[(\frac{1}{5x})(5)] + (\ln 5x)(2) = \dfrac{2x\ln 5x + 2x + 1}{x}$

21. $f'(x) = (1 + \frac{3}{x})\,\dfrac{2x}{x^2+4} + [\ln(x^2+4)](-3x^{-2}) = \dfrac{2x^2(x+3) - 3(x^2+4)\ln(x^2+4)}{x^2(x^2+4)}$

24. $f'(x) = \dfrac{(e^{2x+1})[2(2x+1)^{-1}] - [\ln(2x+1)](2e^{2x+1})}{(e^{2x+1})^2} = \dfrac{2 - 2(2x+1)\ln(2x+1)}{(2x+1)\,e^{2x+1}}$

27. $g(t) = \dfrac{e^t}{t^2+1}$; $g'(t) = \dfrac{(t^2+1)(e^t) - (e^t)(2t)}{(t^2+1)^2} = \dfrac{e^t(t-1)^2}{(t^2+1)^2}$

30. $g'(t) = e^{e^t}\,e^t = e^{e^t+t}$

33. $\dfrac{dy}{dx} = 3(3e^{2x}-1)^2[(3e^{2x})(2)]$

$\qquad = 18e^{2x}(3e^{2x}-1)^2$

36. $\dfrac{dy}{dx} = 5(2x+3\ln x)^4\,(2+ \frac{3}{x}) = \dfrac{5(2x+3)(2x+3\ln x)^4}{x}$

39. $\dfrac{dy}{dx} = (x^2)[3(e^x+2)^2 e^x] + (e^x+2)^3(2x) = x(e^x+2)^2(3xe^x + 2e^x + 4)$

42. $\dfrac{dy}{dx} = 2e^{2x+1}e^{2x+1}(2) = 4e^{2(2x+1)}$

45. (a) $\dfrac{dy}{dt} = 2e^{3t}(3) - 4e^{-t}(-1) = 6e^{3t} + 4e^{-t}$

(b) $\dfrac{dy}{dt} = 3 \dfrac{1}{t^2-4} (2t) = \dfrac{6t}{t^2-4}$

(c) $\dfrac{dy}{dt} = (t^2)(e^t - \tfrac{1}{t}) + (e^t - \ln t)(2t) = t^2 e^t - t + 2t(e^t - \ln t)$

(d) $\dfrac{dy}{dt} = \dfrac{(t^2-5)(2e^{2t}) - (e^{2t})(2t)}{(t^2-5)^2} = \dfrac{2e^{2t}(t^2-t-5)}{(t^2-5)^2}$

(e) $\dfrac{dy}{dt} = e^{(2t+1)^2}[2(2t+1)(2)] = 4(2t+1)e^{(2t+1)^2}$

(f) $\dfrac{dy}{dt} = 4 \dfrac{1}{2t + \sqrt{t}} (2 + \tfrac{1}{2}t^{-1/2}) = \dfrac{2(4\sqrt{t} + 1)}{\sqrt{t}(2t + \sqrt{t})}$

(g) $\dfrac{dy}{dt} = 4[\ln(2t + \sqrt{t})]^3 \dfrac{1}{2t + \sqrt{t}} (2 + \tfrac{1}{2}t^{-1/2}) = \dfrac{2[\ln(2t + \sqrt{t})]^3(4\sqrt{t} + 1)}{\sqrt{t}(2t + \sqrt{t})}$

(h) $\dfrac{dy}{dt} = 6(4t - e^{2/t})^5[4 - e^{2/t}(-2t^{-2})] = \dfrac{12(4t - e^{2/t})^5(2t^2 + e^{2/t})}{t^2}$

48. **Point:** If $x = 1$, $y = (2 + \ln 2)^3 \approx 19.5335$
    Slope: $y' = 3[2 + \ln(1+x)]^2(1+x)^{-1}$ which is $\tfrac{3}{2}(2 + \ln 2)^2 \approx 10.8796$, if
    $\qquad\qquad\qquad\qquad\qquad\qquad\qquad\qquad\qquad\qquad\qquad\quad x = 1$.
    Equation: $y - 19.5335 = 10.8796(x-1)$ or $y = 10.8796x + 8.6539$
    $\qquad\qquad\quad y = [1.5(2+\ln 2)^2]x + (2+\ln 2)^2(0.5+\ln 2)$ is an exact form.

51. (a) $A(t) = 10000e^{0.09t}$; $A(5.5) = 10000e^{0.495} \approx \$16{,}404.98$
    (b) $A'(t) = 900e^{0.09t}$; $A'(10) = 900e^{0.9} \approx \$2{,}213$ per year

    (c) $20000 = 10000e^{0.09t}$; $e^{0.09t} = 2$; $0.09t = \ln 2$; $t = \dfrac{\ln 2}{0.09} \approx 7.70$ yrs.

54. Let $y = f(x)$. $f'(x) = -2(x-1)e^{-(x-1)^2} = 0$, if $x = 1$.
    f is increasing on $(-\infty, 1)$; decreasing on $(1, \infty)$.

    $f''(x) = 2[2(x-1)^2 - 1]e^{-(x-1)^2} = 0$, if $x \approx 0.29$, 1.71
    f is concave up on $(-\infty, 0.29)$ and on $(1.71, \infty)$; down on $(0.29, 1.71)$.

x	f(x)	
1	1	◁ Global maximum
0.29	0.61	◁ Inflection Point
1.71	0.61	◁ Inflection Point
0	0.37	
2	0.37	
-1	0.02	

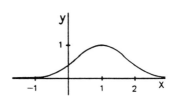

Problem Set 5.4 General Logarithmic and Exponential Functions

3. $e^{2.63\ln4} \approx 38.3193$

6. $\dfrac{\ln 0.098}{\ln 5} \approx -1.4432$

9. $e^{(\ln 4.71)t} \approx e^{1.549879t}$

12. $\dfrac{\ln t}{\ln 3} = [(\ln 3)^{-1}]\ln t \approx 0.098612288\,\ln t$

15. $\dfrac{dy}{dx} = \dfrac{1}{x\ln 5}$

18. $\dfrac{dy}{dx} = 10^x\ln 10$

21. $\dfrac{dy}{dx} = 6^{4-5x^2}(10x\ln 6)$

24. $\dfrac{dy}{dx} = 10\,\dfrac{3x^2-4x}{(x^3-2x^2)\ln 2} = \dfrac{10(3x-4)}{x(x-2)\ln 2}$

27. $\dfrac{dy}{dx} = (2^x)[3^{-2x}(\ln 3)(-2)] + (3^{-2x})(2^x\ln 2) = 2^x3^{-2x}(\ln 2 - 2\ln 3) = 2^x3^{-2x}\ln(\tfrac{2}{9})$

30. $\ln y = \ln(3x+2)^x = x\ln(3x+2)$

$\dfrac{1}{y}\dfrac{dy}{dx} = (x)\left(\dfrac{3}{3x+2}\right) + [\ln(3x+2)](1)$

$\dfrac{dy}{dx} = y\left(\dfrac{3x + (3x+2)\ln(3x+2)}{3x+2}\right) = \dfrac{(3x+2)^x[3x + (3x+2)\ln(3x+2)]}{3x+2}$

$= (3x+2)^{x-1}[3x + (3x+2)\ln(3x+2)]$

33. $\ln y = (x^2-x^{-1})\ln(5x)$

$\dfrac{1}{y}\dfrac{dy}{dx} = (x^2-x^{-1})(\tfrac{5}{5x}) + [\ln(5x)](2x+x^{-2})$

$\dfrac{dy}{dx} = (5x)^{x^2-x^{-1}}[x - x^{-2} + (2x+x^{-2})\ln(5x)]$

36. $\ln y = e^x\ln(2 + 3\ln x)$

$\dfrac{1}{y}\dfrac{dy}{dx} = (e^x)\left(\dfrac{1}{2 + 3\ln x}\,\dfrac{3}{x}\right) + [\ln(2 + 3\ln x)](e^x)$

$\dfrac{dy}{dx} = \dfrac{e^x(2 + 3\ln x)^{e^x}[3 + x(2+3\ln x)\ln(2+3\ln x)]}{x(2+3\ln x)}$

39. $\ln|y| = 3\ln|x+1| + 4\ln|x+2| - 5\ln|x+3|$

$$\frac{1}{y}\frac{dy}{dx} = \frac{3}{x+1} + \frac{4}{x+2} - \frac{5}{x+3}$$

$$\frac{dy}{dx} = y\left(\frac{3}{x+1} + \frac{4}{x+2} - \frac{5}{x+3}\right) = \frac{2(x+1)^2(x+2)^3(x^2+8x+10)}{(x+3)^6}$$

42. $\ln|y| = 4\ln|3x+5| - 3\ln|x^2-x+9|$

$$\frac{1}{y}\frac{dy}{dx} = \frac{12}{3x+5} - \frac{3(2x-1)}{x^2-x+9}$$

$$\frac{dy}{dx} = y\left[\frac{12}{3x+5} - \frac{3(2x-1)}{x^2-x+9}\right] = \frac{-3(3x+5)^3(2x^2+11x-41)}{(x^2-x+9)^4}$$

45. $\dfrac{\ln 100}{\ln 2} \approx 6.6439$; $\dfrac{\ln 100}{\ln e} = \ln 100 \approx 4.6052$; $\log_{10} 100 = 2$; $\dfrac{\ln 100}{\ln 200} \approx 0.8692$

48. **Point:** If $x = 2$, $y = 4$.

Slope: $y' = 4^{x^2-3}(\ln 4)(2x)$ which equals $16\ln 4$ when $x = 2$.

Equation: $y-4 = (16\ln 4)(x-2)$; $y-4 = 22.1807(x-2)$, approximately.

51. $\ln(x\dot{y}^2) = \ln(e^{-5})$; $\ln x + 2\ln y = -5$.

Letting $X = \ln x$ and $Y = \ln y$, we obtain $X+2Y = -5$ or $Y = -\frac{1}{2}X - \frac{5}{2}$, so the slope is $-\frac{1}{2}$ and the Y-intercept is $-\frac{5}{2}$

54. $R = \log_{10}(I/I_0)$, so $10^R = I/I_0$ or $I = I_0 10^R$.

For 1906 San Francisco, $I = I_0 10^{8.2}$; for recent quake, $I = I_0 10^{4.2}$.

$$\frac{I_0 10^{8.2}}{I_0 10^{4.2}} = 10^4 = 10000$$

San Francisco quake of 1906 was 10,000 as intense as the recent one.

Problem Set 5.4

3. (a) 118.9043   (b) 8.611396   (c) 7.550287
  (d) 139.0456   (e) 1.169610   (f) 2.200973

6.

x	f(x)	g(x)
-4	0.09	3.32
-3	0.17	2.46
-2	0.30	1.82
-1	0.55	1.35
0	1	1
1	1.82	0.74
2	3.32	0.55
3	6.05	0.41
4	11.02	0.30

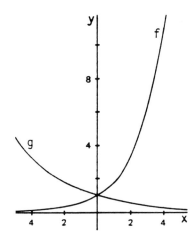

9. $2\ln x + 3\ln 5 - 5\ln y = \ln x^2 + \ln 5^3 - \ln y^5 = \ln\left(\dfrac{125x^2}{y^5}\right)$, $x > 0$

12. (a) $\ln(5x-2) = 3$
$5x-2 = e^3$
$5x = 2+e^3$
$x = \dfrac{2+e^3}{5}$

(b) $3^{2-x} = \pi^2$
$(2-x)\ln 3 = 2\ln\pi$
$x = \dfrac{2\ln 3 - 2\ln\pi}{\ln 3}$

(c) $\ln(2x+1) - \ln(3+4x) = 2$

$\ln\dfrac{2x+1}{4x+3} = 2$

$e^2 = \dfrac{2x+1}{4x+3}$, $x > -1/2$

$4e^2 x + 3e^2 = 2x+1$

$(4e^2-2)x = 1-3e^2$

$x = \dfrac{1-3e^2}{4e^2-2} \approx -0.77$, which is
not greater than $-\frac{1}{2}$, so there
are no solutions.

(d) $e^{2x^2+1} = \ln 2000$

$2x^2+1 = \ln(\ln 2000)$

$x = \pm\sqrt{0.5[\ln(\ln 2000) - 1]}$

$\approx \pm 0.7170$

15. (a) $f'(x) = (1)(e^x) + (x-2)(e^x) = e^x(x-1)$

Values of $f'(x)$ $\underline{\quad\quad (-) \quad\quad (0) \quad\quad (+) \quad\quad}$
$\quad\quad\quad\quad\quad\quad\quad\quad\quad\quad\quad 1 \quad\quad\quad\quad x$

Therefore, f is decreasing on $(-\infty,1)$; increasing on $(1,\infty)$.

(b) $f''(x) = (e^x)(x-1) + (e^x)(1) = xe^x$

Values of $f''(x)$ $\underline{\quad\quad (-) \quad\quad (0) \quad\quad (+) \quad\quad}$
$\quad\quad\quad\quad\quad\quad\quad\quad\quad\quad\quad 0 \quad\quad\quad\quad x$

∴ f is concave down on $(-\infty,0)$; concave up on $(0,\infty)$.

(c) The minimum value of f is $f(1) = -e \approx -2.72$.

(d)

x	f(x)	
-4	$-6/e^4$	≈ -0.11
-3	$-5/e^3$	≈ -0.25
-2	$-4/e^2$	≈ -0.54
-1	$-3/e$	≈ -1.10
0		-2
1	$-e$	≈ -2.72
2		0
3	e^3	≈ 20.09

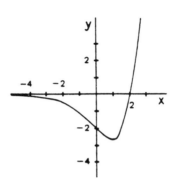

18. $\dfrac{dy}{dx} = \dfrac{2x}{x^2+4}$; $\dfrac{d^2y}{dx^2} = \dfrac{(x^2+4)(2) - (2x)(2x)}{(x^2+4)^2} = \dfrac{8-2x^2}{(x^2+4)^2}$ is 0 if $x = \pm 2$.

Values of $\dfrac{dy}{dx}$ $\underline{\quad (-) \quad (0) \quad (+) \quad (0) \quad (-) \quad}$
$\quad\quad\quad\quad\quad\quad\quad\quad\quad -2 \quad\quad\quad 2 \quad\quad\quad x$

Thus, concavity changes at $x = -2$ and at $x = 2$.

21. $(2\pi)^t = e^{t\ln(2\pi)} = e^{(\ln 2\pi)t} \approx e^{1.837877066t}$

24. $\log_8 t^3 = 3\log_8 t = 3\dfrac{\ln t}{\ln 8} = \dfrac{3}{\ln 8}\ln t \approx 1.442695041\ln t$

27. (a) $\ln f(x) = (x^3+1)\ln(x^2+1)$

$$\frac{1}{f(x)} f'(x) = (3x^2)[\ln(x^2+1)] + (x^3+1)\frac{2x}{x^2+1}$$

$\therefore \frac{1}{4}f'(1) = (3)(\ln 2) + (2)(1)$, so $f'(1) = 4(3\ln 2 + 2) \approx 16.3178$

(b) $\ln f(x) = 6\ln(3x^2-2) + 8\ln(2x-1) - \frac{1}{2}\ln(2x+1)$

$$\frac{1}{f(x)} f'(x) = \frac{6(6x)}{3x^2-2} + \frac{8(2)}{2x-1} - \frac{1}{2}\frac{2}{2x+1}$$

$\therefore \frac{1}{\sqrt{3}} f'(1) = \frac{36}{1} + \frac{16}{1} - \frac{2}{6}$, so $f'(1) = \sqrt{3}(52 - \frac{1}{3}) \approx 89.4893$

30. $\ln y = \ln 3 + (\pi x+2) = \pi x + (2+\ln 3)$, so let $Y = \ln y$.
    Then $Y = \pi x + (2+\ln 3)$; slope is π & Y-intercept is $(2+\ln 3)$.

33. $\frac{dy}{dt} = 6e^{x^2/4}\frac{2x}{4}\frac{dx}{dt}$, so when $x = 2$, $\frac{dy}{dt} = 6e(1)(3) = 18e \approx 48.9291$

36. It is $f'(2)$ if $f(x) = e^{3x^2}$.

    $f'(x) = e^{3x^2}(6x)$, so $f'(2) = 12e^{12} \approx 1953057.497$

39. (a) $f(x) = (\ln x)(\ln x-1)$ is 0 if $x = 1$ or $x = e$.

    Values of $f(x)$ $\quad\underline{\quad (+) \quad (0) \quad (-) \quad (0) \quad (+)\quad}$
    $\qquad\qquad\qquad\qquad 0 \qquad\quad 1 \qquad\quad e \qquad\quad x$

    Therefore, $f(x) > 0$ on $(0,1) \cup (e,\infty) \approx (0,1) \cup (2.72,\infty)$.

    (b) $f'(x) = \frac{2(\ln x)}{x} - \frac{1}{x} = \frac{2\ln x-1}{x}$ is 0 if $x = e^{1/2}$.

    Values of $f'(x)$ $\quad\underline{\quad (-) \quad (0) \quad (+)\quad}$
    $\qquad\qquad\qquad\qquad 0 \qquad\ e^{1/2} \qquad x$

    Therefore, $f'(x) > 0$ on $(\sqrt{e},\infty) \approx (1.65,\infty)$.

    (c) $f''(x) = \frac{(x)(2/x) - (2\ln x-1)(1)}{x^2} = \frac{3-2\ln x}{x^2}$ is 0 if $x = e^{3/2}$.

    Values of $f''(x)$ $\quad\underline{\quad (+) \quad (0) \quad (-)\quad}$
    $\qquad\qquad\qquad\qquad 0 \qquad\ e^{3/2} \qquad x$

    Therefore, $f''(x) > 0$ on $(0,e^{3/2}) \approx (0,4.48)$.

Chapter 5 Review Problem Set

Problem Set 6.1   Exponential Growth and Decay

3. $y = 16e^{0.8t}$ $(k = 0.8, y_0 = 16)$; If $t = 8$, $y = 16e^{6.4} \approx 9630$

6. $y = 40e^{-0.006t}$ $(k = -0.006, y_0 = 40)$; If $t = 8$, $y = 40e^{-0.048} \approx 38.13$

9. Let y be the number of members after t minutes.   Then $y = 100e^{kt}$.

   (a) Doubling will take place three times: 100 to 200 to 400 to 800;

   (b) $y = 200$ when $t = 70$, so $200 = 100e^{70k}$; $k = \dfrac{\ln 2}{70} \approx 0.009902$.

   Therefore, $y = 100e^{0.009902t}$ (approximately).

   (c) When $t = 120$, $y = 100e^{0.009902(120)} \approx 328$ members.

12. (a) Let y be the number of members after t hours.   Then $y = 3500e^{kt}$.

   $y = 800$ when $t = -6$, so $800 = 3500e^{-6k}$; $k = \dfrac{\ln 4.375}{6} \approx 0.2460$.

   Therefore, $y = 3500e^{0.2460t}$ (approximately).

   Then, when $t = 4$, $y = 3500e^{0.2460(4)} \approx 9363$ members.

   (b) If $y = 10000$, $10000 = 3500e^{0.2460t}$; $t = \dfrac{\ln(20/7)}{0.2460} \approx 4.27$ hrs.

15. (a) Let y be the population in millions t years after 1/1/78.

   Then $y = 10.5e^{kt}$.

   $y = 14.6$ if $t = 10$; $14.6 = 10.5e^{10k}$; $k = \dfrac{\ln(14.6/10.5)}{10} \approx 0.03296$.

   Therefore, $y = 10.5e^{0.03296t}$ (approximately).

   (b) When $t = 22$ (Jan. 1, 2000), $y = 10.5e^{0.03296(22)} \approx 21.7$ million.

18. Let y be the number of grams left after t seconds. Then $y = 5e^{kt}$.

$y = 2.5$ when $t = 3.92$, so $2.5 = 5e^{3.92k}$; $k = \dfrac{ln(0.5)}{3.92} \approx -0.1768$.

Therefore, $y = 5e^{-0.1768t}$ (approximately).

When $t = 2.5$, $y = 5e^{-0.1768(2.5)} \approx 3.2$ grams.

21. $y = e^{-0.000124443t}$ (approximately) [See Example E.]

If $y = 0.62$, then $0.62 = e^{-0.000124443t}$; $t = \dfrac{ln(0.62)}{-0.000124443} \approx 3,840$ yrs.

24. $y = e^{-0.000124443t}$ (approximately) [See Example E.]

If $y = 0.71$, then $0.71 = e^{-0.000124443t}$; $t = \dfrac{ln(0.71)}{-0.000124443} \approx 2,750$ yrs.

If $y = 0.75$, then $0.75 = e^{-0.000124443t}$; $t = \dfrac{ln(0.75)}{-0.000124443} \approx 2,310$ yrs.

The burial site had been used about $2,750 - 2,310 = 440$ years.

27. It was halved twice, each time taking about 5570 years, so it is about 11,140 years old.

30. Let y be the population in thousands t years after 1975.

Then $y = 48e^{kt}$.

$y = 62$ when $t = 10$, so $62 = 48e^{10k}$; $k = \dfrac{ln(31/24)}{10} \approx 0.02559$.

Therefore, $y = 48e^{0.02559t}$ (approximately).

Then, when $t = 30$, $y = 48e^{0.02559(30)} \approx 103$ thousand.

33. Let N(A) and N(B) be the populations of Colonies A & B, respectively, t hours from now.

Then $N(A) = N_0(A)e^{0.4t}$ and $N(B) = N_0(B)e^{1.2t}$.

$N_0(A) = 10N_0(B)$, so $N(A) = 10N_0(B)e^{0.4t}$.

Thus, $N(A) = N(B)$ if $10N_0(B)e^{0.4t} = N_0(B)e^{1.2t}$, so $t = \dfrac{ln10}{0.8} \approx 2.9$ hrs.

3. [Simple Bounded Growth model with L = 20, k = 0.45, y_0 = 12]
   $y = 20-(20-12)e^{-0.45t} = 20-8e^{-0.45t}$
   When t = 4, $y = 20-8e^{-1.8} \approx 18.7$

6. [Simple Bounded Growth model with L = 1 (100%), y_0 = 0] $y = 1(1-e^{-kt})$

   (a) If t = 10 & y = 0.40, $0.40 = 1-e^{-10k}$; $k = \dfrac{\ln(0.6)}{-10} \approx 0.05108$

   Thus, $y = 1-e^{-0.05108t}$ (approximately), so about $2400(1-e^{-0.05108t})$ viewers will have seen the advertisement within t days.

   (b) If y = 0.80 (80%), $0.80 = 1-e^{-0.05108t}$, so $t = \dfrac{\ln(0.2)}{-0.05108} \approx 31.5$ days.

9. [Simple Bounded Growth model with L = 72, W_0 = 0, k = 0.6]

   (a) $W = 72(1-e^{-0.6t})$
   (b) 72 words per minute   (W approaches 72 as t goes to infinity.)

   (c) If W = 68, $68 = 72(1-e^{-0.6t})$; $t = \dfrac{\ln 18}{0.6} \approx 4.8$ weeks.

12. [Simple Bounded Growth model with L = 20, y_0 = 1]
    $y = 20-(20-1)e^{-kt} = 20-19e^{-kt}$
    If t = 1 and y = 6, $6 = 20-19e^{-k}$; $k = -\ln(14/19) \approx 0.3054$
    Therefore, $y = 20-19e^{-0.3054t}$ (approximately).

    (a) If t = 4, $y = 20-19e^{-0.3054(4)} = 20-19e^{-1.2216} \approx 14.4$ cm.

    (b) If y = 18, $18 = 20-19e^{-0.3054t}$, so $t = \dfrac{\ln(2/19)}{-0.3054} \approx 7.4$ months.

15. Let y be the number of thousands of people infected after t days.

    $L = 120$, $y_0 = 0.03$, $B = \dfrac{120 - 0.03}{0.03} = 3999$

    $y = \dfrac{120}{1 + 3999e^{-120kt}}$

    If t = 10 & y = 3, $3 = \dfrac{120}{1 + 3999e^{-1200k}}$; so $k = \dfrac{\ln(13/1333)}{-1200} \approx 0.003859$

    Therefore, $y = \dfrac{120}{1 + 3999e^{-0.4631t}}$ (approximately).

    If y = 100, $100 = \dfrac{120}{1 + 3999e^{-0.4631t}}$; $t = \dfrac{\ln 19995}{0.4631} \approx 21.4$ days.

18. [Simple Bounded Growth model with $L = 40$, $P_0 = 0$]

$P = 40(1-e^{-kt})$

If $t = 5$ and $P = 25$, $25 = 40(1-e^{-5k})$; $k = \dfrac{ln(3/8)}{-5} \approx 0.1962$

Therefore, $P = 40(1-e^{-0.1962t})$ (approximately).

Then, if $t = 8$, $P = 40(1-e^{-1.5696}) \approx 32$ sets.

21. Let y be the number of thousands of homes after t years.

$L = 20$, $y_0 = 0.1$, $B = \dfrac{20 - 0.1}{0.1} = 199$

$y = \dfrac{20}{1 + 199e^{-20kt}}$

If $t = 5$ & $y = 0.8$, $0.8 = \dfrac{20}{1 + 199e^{-1200k}}$; so $k = \dfrac{ln(192/1592)}{-100} \approx 0.02115$

Therefore, $y = \dfrac{20}{1 + 199e^{-0.423t}}$

If $t = 20$, $y = \dfrac{20}{1 + 199e^{-8.46}} \approx 19.2$

19.2 thousand homes with an average of 1.2 school-aged children means a school system with about 23,000 pupils.

24. $P(x) = 5.5[480 - 380e^{-0.003209x}] - 0.80x$

$P'(x) = 5.5[1.21942e^{-0.003209x}] - 0.8 = 0$,

if $x = \dfrac{ln(0.8/6.70681)}{-0.003209} \approx 663$ pounds per acre.

Problem Set 6.3 Mathematics of Finance

3. $F = P + Prt = P(1 + rt) = 1200[1 + (0.095)(4.5)] = \$1,713$

6. $F = P + Prt = P(1 + rt) = 2182[1 + (0.0821)(14.5)] = \$4,779.56$

9. $F = P(1 + r)^t = 1250(1 + 0.095)^{11} = \$3,392.07$

12. (a) $F = 1000(1 + 0.08)^{10} = \$2,158.92$

(b) $F = 1000(1 + \frac{0.08}{2})^{20} = \$2,191.12$

(c) $F = 1000(1 + \frac{0.08}{4})^{40} = \$2,208.04$

(d) $F = 1000(1 + \frac{0.08}{12})^{120} = \$2,219.64$

(e) $F = 1000(1 + \frac{0.08}{365})^{3650} = \$2,225.35$

15. $F = 1000e^{(0.09)(10)} = \$2,459.60$ **18.** $F = 2182e^{(0.0821)(14.5)} = \$7,175.64$

21. $r_e = (1 + \frac{0.089}{365})^{365} - 1 \approx 0.093069 \ (9.3069\%)$

24. 10% compounded quarterly: $r_e = (1 + \frac{0.10}{4})^4 - 1 \approx 0.1038 \ (10.38\%)$

9.9% compounded quarterly: $r_e = (1 + \frac{0.099}{12})^{12} - 1 \approx 0.1036 \ (10.36\%)$

The 10% compounded quarterly gives a slightly better return.

27. Invest \$1 at 11% compounded monthly. Let $F = \$2$ and solve for t.

$2 = 1(1 + \frac{0.11}{12})^{12t}$; $t = \dfrac{ln2}{12ln(1 + 0.11/12)} \approx 6.33$ years   (76 months)

30. $3 = 1(1 + \frac{0.0636}{365})^{365t}$; $t = \dfrac{ln3}{365ln(1 + 0.0636/365)} \approx 17.2753$ years

It would take 17 years and 101 days. (In fact, it would take 4 or 5 days less due to leap years.)

33. $4800 = P(1 + \frac{0.084}{4})^{48}$, so $P = 4800(1 + \frac{0.084}{4})^{-48} = \$1,770.13$

36. $4800 = 1000e^{0.10t}$, so $t = 10ln4.8 \approx 15.68615918$ years which is about 15 yrs, 250 days (ignoring leap years), 10 hrs, 45 mins, 16 secs.

39. $3000 = 1000e^{15r}$, so $r = \dfrac{ln3}{15} \approx 0.073240819$ (7.3240819%)

42. $F_4 = 1000(1 + \frac{0.10}{12})^{48}$ at end of 4 years. [F_n is amount at end of n yrs]

$F_8 = [1000(1 + \frac{0.10}{12})^{48}](1 + \frac{0.105}{12})^{48} = 1000[(1 + \frac{0.10}{12})(1 + \frac{0.105}{12})]^{48}$

$F_{12} = 1000[(1 + \frac{0.10}{12})(1 + \frac{0.105}{12})(1 + \frac{0.11}{12})]^{48}$

$F_{16} = 1000[(1 + \frac{0.10}{12})(1 + \frac{0.105}{12})(1 + \frac{0.11}{12})(1 + \frac{0.115}{12})]^{48}$

$F_{18} = 1000[(1 + \frac{0.10}{12})(1 + \frac{0.105}{12})(1 + \frac{0.11}{12})(1 + \frac{0.115}{12})]^{48}(1 + \frac{0.12}{12})^2$

$= \$7,036.63$

3. $2500(1 + \frac{0.08}{12})^{-120} = \$1,126.31$ 6. $5000(1 + \frac{0.10}{1})^{-12} = \$1,593.15$

9. $2000(1 + \frac{0.09}{1})^{-2} + 3500(1 + \frac{0.09}{1})^{-6} = \$3,770.30$

12. $2400[(1 + \frac{0.09}{1})^{-2} + (1 + \frac{0.09}{1})^{-4} + (1 + \frac{0.09}{1})^{-6}] = \$5,151.29$

15. $6000(1.08)^{-1} = 3000 + x(1.08)^{-5}$

$x = [6000(1.08)^{-1} - 3000](1.08)^5 = \$3,754.95$

$$\begin{array}{ccc} \dfrac{6000}{1} & & t \\ \hline \dfrac{3000}{0} & \dfrac{x}{5} & t \end{array}$$

18. $8000 = x(1 + \frac{0.06}{4})^{-8} + 2x(1 + \frac{0.06}{4})^{-16}$

$x = \dfrac{8000}{(1.015)^{-8} + 2(1.015)^{-16}} = \$3,247.05$

$$\begin{array}{ccc} \dfrac{8000}{0} & & t \\ & \dfrac{x}{2} \quad \dfrac{2x}{4} & t \end{array}$$

21. The true rates are $\dfrac{0.08 - 0.04}{1 + 0.04} \approx 0.038461538$

and $\dfrac{0.14 - 1.10}{1 + 0.10} \approx 0.036363636$,

so the 8% interest with 4% inflation is better for the investor.

24. $1 - \frac{2}{12} = \frac{5}{6}$

(a) $V = 80000(\frac{5}{6})^4 = \$38,580.25$ (b) $V = 80000(\frac{5}{6})^8 = \$18,605.44$

(c) $V = 80000(\frac{5}{6})^{12} = \$8,972.53$

27. $1 + \frac{0.09}{12} = 1.0075$

$$\text{old} \quad \begin{array}{ccc} \dfrac{2000}{-4} & \dfrac{4000}{-1} & t \end{array} \qquad \text{new} \quad \begin{array}{ccc} \dfrac{x}{-3} & \dfrac{x}{0} & t \end{array}$$

$2000(1.0075)^{48} + 4000(1.0075)^{12} = x(1.0075)^{36} + x$

$x = \dfrac{2000(1.0075)^{48} + 4000(1.0075)^{12}}{(1.0075)^{36} + 1} \approx \$3,135.19$

30. old $\dfrac{\quad\quad x \quad\quad}{-4 \qquad\qquad\qquad\qquad\qquad t}$ new $\dfrac{10000 \;\; 10000 \;\; 10000 \;\; 10000}{-3 \qquad -2 \qquad -1 \qquad 0 \;\; t}$

$x(1.06)^4 = 10000[(1.06)^3 + (1.06)^2 + (1.06)^1 + 1]$

$x = \dfrac{10000[(1.06)^3 + (1.06)^2 + (1.06)^1 + 1]}{(1.06)^4} \approx \$34,651.06$

33. $10000 = 25000(1+x)^{-10}$, so $x = \sqrt[10]{2.5} - 1 \approx 0.095958226 = 9.5958226\%$

Chapter 6 Review Problem Set

3. [Simple bounded growth model with $k = 0.07$, $L = 12$, $y_0 = 3$.]

$y = 12 - (12-3)e^{-0.07t} = 12 - 9e^{-0.07t}$

If $t = 5$, then $y = 12 - 9e^{-0.07(5)} \approx 5.66$

6. [Logistic growth model with $k = 0.04$, $L = 30$, $y_0 = 10$.]

$B = 2$ and $Lk = 1.2$, so $y = \dfrac{30}{1+2e^{-1.2t}}$

If $t = 6$, then $y = \dfrac{30}{1+2e^{-7.2}} \approx 29.96$

9. $y = e^{-0.000124443t}$, so we need to solve $0.06 = e^{-0.000124443t}$ for t.

$t \approx \dfrac{\ln(0.06)}{-0.000124443} \approx 22608$, so it is about 22,600 years old.

12. $L = 84000$, $y_0 = 1500$, so $y = 84000 - (84000-1500)e^{-kt}$

$y = 63000$ when $t = 3$, so $63000 = 84000 - 82500e^{-3k}$

$\therefore k = -\frac{1}{3}\ln(\frac{21000}{82500}) \approx 0.45609195$

At 4:00 p.m., $t = 7$, so $y = 84000 - 82500e^{-(0.45609295)(7)} \approx 80,612$

15. (a) $3000[1 + (0.08)(10)]$ — $\$5,400.00$

(b) $3000(1 + 0.08)^{10}$ — $\$6,476.77$

(c) $3000(1 + \frac{0.08}{12})^{120}$ — $\$6,658.92$

(d) $3000e^{(0.08)(10)}$ — $\$6,676.62$

18. (a) $5000 = 3200(1 + \frac{0.095}{12})^n$

$(1 + \frac{0.095}{12})^n = 1.5625$

$n\ln(1 + \frac{0.095}{12}) = \ln(1.5625)$, so $n \approx 56.6$

It will take 57 months.

(b) $5000 = 3200e^{0.095t}$; $e^{0.095t} = 1.5625$; $0.095t = \ln(1.5625)$

∴ $t \approx 4.697758975$ years [4 yr, 8 mo, 11 da, 4 hr, 38 min, 15 sec]

21. $P = 560(1.1)^{-1} + 560(1.1)^{-2} + 560(1.1)^{-3} + 560(1.1)^{-4} = \$1,775.12$

24. $10000 = 4500(1+i)^{20}$; $(1+i)^{20} = \frac{100}{45}$

∴ $i = (\frac{100}{45})^{1/20} - 1 \approx 0.040733116$ per period,

so $2(0.040733116) \approx 8.147\%$ is realized.

27. $1 - \frac{2}{10} = 0.8$

(a) $V = 120000(0.8)^5 = \$39,321.60$

(b) $60000 = 120000(0.8)^n$; $(0.8)^n = 0.5$; $n = \frac{\ln(0.5)}{\ln(0.8)} \approx 3.106$

It will fall below $\$60,000$ in about 3 years, 39 days.

30. [Simple growth model with $y_0 = 1$] $y = e^{kt}$

$y = 0.67$ when $t = 10$, so $0.67 = e^{10k}$; then $k = \frac{\ln(0.67)}{10} \approx -0.040047756$

If $t = 500$, then $y = e^{(-0.040047756)(500)} \approx 0$ [Virtually none left.]

33. $\dfrac{dy}{dt} = -(L-y_0)e^{-kt}(-k) = k(L-y_0)e^{-kt}$

$K(L-y) = k\{L - [L - (L-y_0)]e^{-kt}\} = k(L-y_0)e^{-kt}$

Therefore, $\dfrac{dy}{dt} = k(L-y)$.

36. $y' = Ce^{3t} + (B+Ct)(3e^{3t}) = (3B+C)e^{3t} + 3Cte^{3t}$

$y'' = 3(3B+C)e^{3t} + [(3C)(e^{3t}) + (3Ct)(3e^{3t})] = (9B+6C)e^{3t} + 9Cte^{3t}$

$y''' = 3(9B+6C)e^{3t} + [(9C)(e^{3t}) + (9Ct)(3e^{3t})] = (27B+27C)e^{3t} + 27Cte^{3t}$

$\therefore\ y''' - 6y'' + 9y' = [(27B+27C)e^{3t} + 27Cte^{3t}]$

$$- 6[(9B+6C)e^{3t} + 9Cte^{3t}]$$

$$+ 9[(3B+C)e^{3t} + 3Cte^{3t}]$$

$$= 0$$

39. $\dfrac{d^2y}{dt^2} = k\left(\dfrac{dy}{dt}\right)^2$

Problem Set 7.1   Antiderivatives

3. f(x)

6. $\frac{x^5}{5}$ + C

9. $\frac{x^{-2}}{-2}$ + C

12. $\frac{x^{7/4}}{7/4}$ + C

15. f(x) = x^{-4} , so the general antiderivative of f(x) is $\frac{x^{-3}}{-3}$ + C.

18. f(x) = x$^{7/2}$, so the general antiderivative of f(x) is $\frac{x^{9/2}}{9/2}$ + C.

21. $5\int x^4 dx - 6\int x dx = 5\frac{x^5}{5} - 6\frac{x^2}{2} + C = x^5 - 3x^2 + C$

24. $4\int x^{1/3} dx + 3\int 1 dx = 4\frac{x^{4/3}}{4/3} + 3x + C = 3x^{4/3} + 3x + C$

27. $5\int x^{-2} dx - 2\int x^{-3} dx = 5\frac{x^{-1}}{-1} - 2\frac{x^{-2}}{-2} + C = -5x^{-1} + x^{-2} + C$

30. $\int x^{-2} dx + 4\int e^x dx = \frac{x^{-1}}{-1} + 4e^x + C = -x^{-1} + 4e^x + C$

33. $2\int u^3 du - 4\int u^{-2} du = 2\frac{u^4}{4} - 4\frac{u^{-1}}{-1} + C = \frac{1}{2}u^4 + 4u^{-1} + C$

36. $\int (3x^2 + 5x - 2) dx = x^3 + \frac{5}{2}x^2 - 2x + C$

39. $\int (1 + 4x^{-1} + 2x^{-2}) dx = x + 4\ell n|x| - 2x^{-1} + C$

42. $F(x) = \int (3-2x) dx = 3x - x^2 + C$, for some C.
    $7 = F(2) = 3(2) - (2)^2 + C$ infers C = 5, so F(x) = 3x - x^2 + 5.

45. $F(x) = \int 4e^x dx = 4e^x + C$, for some C.

   $2 = F(0) = 4e^0 + C$ infers $C = -2$, so $F(x) = 4e^x - 2$.

48. $F(x) = \int (x^2 - 2x^{-1}) dx = \frac{1}{3}x^3 - 2\ell n|x| + C$, for some C.

   $\frac{10}{3} = F(1) = \frac{1}{3}(1)^3 - 2\ell n|1| + C$ infers $C = 3$, so $F(x) = \frac{1}{3}x^3 - 2\ell n|x| + 3$.

51. $R'(x) = 3000 - 0.9x$ (Marginal revenue)
   Therefore, $R(x) = 3000x - 0.45x^2 + C$, for some C.
   $0 = R(0) = 3000(0) - 0.45(0)^2 + C$ infers $C = 0$, so $R(x) = 3000x - 0.45x^2$.
   Then, $R(1500) = 3000(1500) - 0.45(1500)^2 = \$3,487,500$.

54. $a(t) = 8$

   $v(t) = 8t + C_1$, for some C_1. $v = 0$ when $t = 0$ infers $C_1 = 0$,
     so $v(t) = 8t$. Therefore, $v(5) = 40$ ft/sec.

   $s(t) = 4t^2 + C_2$, for some C_2. $s = 0$ when $t = 0$ infers $C_2 = 0$,
     so $s(t) = 4t^2$. Therefore, $s(5) = 100$ ft.

57. $C'(x) = 400x^{-1/2}$ (Marginal cost)
   Therefore, $C(x) = 800x^{1/2} + K$, for some K.
   $150000 = C(0) = 0 + K$ infers $K = 150000$, so $C(x) = 800x^{1/2} + 150000$.

   (a) $C(25600) = 800(25600)^{1/2} + 150000 = \$278,000$

   (b) $\dfrac{278,000}{25600} = \$10.86$ per book.

Problem Set 7.2   The Substitution Rule

3. $\int (x^2+3) 2x dx = \int u^2 du$

   $= \frac{1}{3}u^3 + C_1 = \frac{1}{3}(x^2+3)^3 + C_1$

   $\int (x^2+3) 2x dx = \int (2x^5 + 12x^3 + 18x) dx = \frac{1}{3}x^6 + 3x^4 + 9x^2 + C_2$

   [These two forms are equivalent, with $C_2 = C_1 + 9$.]

> Let $u = x^2 + 3$
> Then $du = 2x dx$

6. $\int u^3 du = \frac{1}{4}u^4 + C = \frac{1}{4}(2x^3+1)^4 + C$

> Let $u = 2x^3 + 1$
> Then $du = 6x dx$

9. $\int u^4 du = \frac{1}{5}u^5 + C = \frac{1}{5}(x^2+6x+4)^5 + C$

$$\boxed{\begin{array}{l} \text{Let } u = x^2+6x+4 \\ \text{Then } du = (2x+6)dx \end{array}}$$

12. $\frac{1}{3}\int (x^3+3x^2-11)^{11} \, 3(x^2+2x)dx = \frac{1}{3}\int u^7 du$

$\qquad = \frac{1}{24}u^8 + C = \frac{1}{24}(x^3+3x^2-11)^8 + C$

$$\boxed{\begin{array}{l} \text{Let } u = x^3+3x^2-11 \\ \text{Then } du = (3x^2+6x)dx = 3(x^2+2x)dx \end{array}}$$

15. $\frac{1}{3}\int e^{3x} \, 3dx = \frac{1}{3}\int e^u du = \frac{1}{3}e^u + C = \frac{1}{3}e^{3x} + C$

$$\boxed{\begin{array}{l} \text{Let } u = 3x \\ \text{Then } du = 3dx \end{array}}$$

18. $\frac{4}{3}\int e^{3u-4} \, 3du = \frac{4}{3}\int e^z dz = \frac{4}{3}e^z + C = \frac{4}{3}e^{3u-4} + C$

$$\boxed{\begin{array}{l} \text{Let } z = 3u-4 \\ \text{Then } dz = 3du \end{array}}$$

21. $2\int e^{\sqrt{x}} \, (\frac{1}{2}x^{-1/2}) \, dx = 2\int e^u du = 2e^u + C = 2e^{\sqrt{x}} + C$

$$\boxed{\begin{array}{l} \text{Let } u = \sqrt{x} \\ \text{Then } du = \frac{1}{2}x^{-1/2}dx \end{array}}$$

24. $\int \frac{1}{u} \, du = \ln|u| + C = \ln|x^3-x+7| + C$

$$\boxed{\begin{array}{l} \text{Let } u = x^3-x+7 \\ \text{Then } du = (3x^2-1)dx \end{array}}$$

27. $\frac{1}{3}\int \frac{1}{u} \, du = \frac{1}{3}\ln|u| + C = \frac{1}{3}\ln(3e^x+1) + C$

$$\boxed{\begin{array}{l} \text{Let } u = 3e^x+1 \\ \text{Then } du = 3e^x dx \end{array}}$$

30. $\frac{1}{3}\int (2+3\ln x)^{1/2}(\frac{3}{x}) \, dx = \frac{1}{3}\int u^{1/2} du = \frac{2}{9}u^{3/2} + C$

$\qquad = \frac{2}{9}(2+3\ln x)^{3/2} + C$

$$\boxed{\begin{array}{l} \text{Let } u = 2+3\ln x \\ \text{Then } du = \frac{3}{x}dx \end{array}}$$

33. $\frac{1}{2}\int \ln(x^2+1) \, \frac{2x}{x^2+1} \, dx = \frac{1}{2}\int u \, du = \frac{1}{4}u^2 + C$

$\qquad = \frac{1}{4}[\ln(x^2+1)]^2 + C$

$$\boxed{\begin{array}{l} \text{Let } u = \ln(x^2+1) \\ \text{Then } du = \dfrac{2x}{x^2+1} \, dx \end{array}}$$

36. $\int u^{10}(u+2)\,du = \int (u^{11}+2u^{10})\,du = \frac{1}{12}u^{12} + \frac{2}{11}u^{11} + C$ $\boxed{\begin{array}{l}\text{Let } u = x-2; \text{ so } x = u+2 \\ \text{Then } du = dx\end{array}}$

$\qquad = \frac{1}{12}(x-2)^{12} + \frac{2}{11}(x-2)^{11} + C$

39. $\int (x^4+10x^2+25)\,dx = \frac{1}{5}x^5 + \frac{10}{3}x^3 + 25x + C$

42. $\int (9x^{-2}+12x^{-1}+4)\,dx = -9x^{-1}+12\ell n|x|+4x+C$

45. $\int (2+5x^{-1/2})\,dx = 2x+10x^{1/2}+C$

48. $\int [x+3-10(x+2)^{-1}]\,dx = \frac{1}{2}x^2+3x-10\ell n|x+2|+C$

51. $\int (8u^6+12u^4+6u^2+1)\,du = \frac{8}{7}u^7 + \frac{12}{5}u^5+2u^3+u+C$

54. $\int (e^{4x}-6e^{2x}+9)\,dx = \frac{1}{4}e^{4x}-3e^{2x}+9x+C$

57. $\frac{1}{3}\int \frac{3(t^2+1)}{t^3+3t-4}\,dt = \frac{1}{3}\int \frac{1}{u}\,du = \frac{1}{3}\ell n|u| + C$ $\boxed{\begin{array}{l}\text{Let } u = t^3+3t-4 \\ \text{Then } du = (3t^2+3)dt = 3(t^3+1)dt\end{array}}$

$\qquad = \frac{1}{3}\ell n|t^3+3t-4| + C$

60. $\int [t+6-18(t+3)^{-1}]\,dt = \frac{1}{2}t^2+6t-18\ell n|t+3|+C$

63. Let $f'(x) = \frac{2x+3}{x^2+3x+1}$ be the slope function. $f(x)$ is an antiderivative.

$\int \frac{2x+3}{x^2+3x+1}\,dx = \int \frac{1}{u}\,du = \ell n|u|+C = \ell n|x^2+3x+1|+C,$ $\boxed{\begin{array}{l}\text{Let } u = x^2+3x+1 \\ \text{Then } du = (2x+3)dx\end{array}}$
so $f(x) = \ell n|x^2+3x+1|+C$, for some C.
The graph passes through $(0,3)$, so $3 = \ell n|(0)^2+3(0)+1|+C = C.$
Therefore, $f(x) = \ell n|x^2+3x+1|+3.$

66. Let $B'(t) = \frac{-1500t^2}{1+t^3}.$ $B(t)$ is one of the antiderivatives of $B'(t)$.

$\int \frac{-1500t^2}{1+t^3}\,dt = -500\int \frac{1}{u}\,du = -500\ell nu + C$ $\boxed{\begin{array}{l}\text{Let } u = 1+t^3 \\ \text{Then } du = 3t^2 dt\end{array}}$

$\qquad = -500\ell n(1+t^3) + C.$ Thus, $B(t) = -500\ell n(1+t^3) + C$, for some C.

$B = 6000$ when $t = 0$, so $C = 6000.$ Then $B(t) = -500\ell n(1+t^3) + 6000.$
Therefore, $B(8) = -500\ell n[1+(8)^3] + 6000 \approx 2880$ bacteria per liter.

3. $\int x \ell nx = (\ell nx)(\frac{1}{2}x^2) - \int (\frac{1}{2}x^2)(x^{-1}) dx$

$= \frac{1}{2}x^2 \ell nx - \frac{1}{2}\int x dx$

$= \frac{1}{2}x^2 \ell nx - \frac{1}{4}x^2 + C = \frac{1}{4}x^2(2\ell nx - 1) + C$

> Let u = ℓnx & dv = xdx
> Then du = $x^{-1}dx$ & v = $\frac{1}{2}x^2$

6. $\frac{2}{9}x^{9/2}\ell nx - \frac{2}{9}\int x^{7/2} dx$

$= \frac{2}{9}x^{9/2}\ell nx - \frac{4}{81}x^{9/2} + C$

$= \frac{2}{81}x^{9/2}(9\ell nx - 2) + C$

> Letting u = ℓnx & dv = $x^{7/2}dx$
> Then du = $x^{-1}dx$ & v = $\frac{2}{9}x^{9/2}$

9. $-\frac{1}{2}x^2 e^{-2x} - \int -x e^{-2x} dx$

$= -\frac{1}{2}x^2 e^{-2x} - [\frac{1}{2}x e^{-2x} - \int \frac{1}{2}e^{-2x}dx]$

$= -\frac{1}{2}x^2 e^{-2x} - \frac{1}{2}x e^{-2x} - \frac{1}{4}e^{-2x} + C$

$= -\frac{1}{4}e^{-2x}(2x^2+2x+1) + C$

> Letting u = x^2 & dv = $e^{-2x}dx$
> Then du = 2xdx & v = $-\frac{1}{2}e^{-2x}$

> Letting u = x & dv = $-e^{-2x}dx$
> Then du = dx & v = $\frac{1}{2}e^{-2x}$

12. $(x-2)\ell n(x-2) - \int 1 dx$

$= (x-2)\ell n(x-2) - x + C$

> Letting u = $\ell n(x-2)$ & dv = dx
> Then du = $(x-2)^{-1}dx$ & v = x-2
> [Note that v = x-2 is a better
> choice for an antiderivative
> for dv than v = x.]

15. $\frac{5x-9}{x(x-3)} = \frac{A}{x} + \frac{B}{x-3} = \frac{A(x-3) + Bx}{x(x-3)}$. Hence, $5x-9 = A(x-3) + Bx$.

Therefore, $\frac{5x-9}{x^2-3x} = \frac{3}{x} + \frac{2}{x-3}$.

> Let x = 0; obtain A = 3.
> Let x = 3; obtain B = 2.

18. $\frac{6x^2-8x-4}{(x-2)(x-1)(x+2)} = \frac{A}{x-2} + \frac{B}{x-1} + \frac{C}{x+2} = \frac{A(x-1)(x+2)+B(x-2)(x+2)+C(x-2)(x-1)}{(x-2)(x-1)(x+2)}$

Therefore, $\frac{6x^2-8x-4}{(x-2)(x-1)(x+2)} = \frac{1}{x-2} + \frac{2}{x-1} + \frac{3}{x+2}$.

> x = 2 infers A = 1
> x = 1 infers B = 2
> x = -2 infers C = 3

21. $\int\left(\dfrac{3}{x+1} - \dfrac{3}{x+2}\right)dx = 3\ell n|x+1| - 3\ell n|x+2| + C$

24. $\int\left(\dfrac{4}{x-1} - \dfrac{3}{x+2}\right)dx = 4\ell n|x-1| - 3\ell n|x+2| + C$

27. $\int\left(\dfrac{5}{x} + \dfrac{2}{x+3} - \dfrac{4}{x-2}\right)dx = 5\ell n|x| + 2\ell n|x+3| - 4\ell n|x-2| + C$

30. $-2e^{-3x} + C$ [Use substitution $u = e^{-3x}$, $du = -3e^{-3x}dx$, if necessary.]

33. $\dfrac{1}{8}\int e^{u}\,du = \dfrac{1}{8}e^{u} + C = \dfrac{1}{8}e^{4x^2} + C$

> Let $u = 4x^2$
> Then $du = 8xdx$

36. $\int u^{4}\,du = \dfrac{1}{5}u^{5} + C = \dfrac{1}{5}(\ell nx)^{5} + C$

> Let $u = \ell nx$
> Then $du = x^{-1}dx$

39. $\dfrac{2x+7}{(x+5)(x-2)} = \dfrac{A}{x+5} + \dfrac{B}{x-2} = \dfrac{A(x-2) + B(x+5)}{(x+5)(x-2)}$, so $2x+7 = A(x-2) + B(x+5)$.

Therefore, $\dfrac{2x+7}{x^2+3x-10} = \dfrac{3/7}{x+5} + \dfrac{11/7}{x-2}$.

> Let $x = -5$; obtain $A = 3/7$.
> Let $x = 2$; obtain $B = 11/7$.

$\int\left(\dfrac{3/7}{x+5} + \dfrac{11/7}{x-2}\right)dx = (3/7)\ell n|x+5| + (11/7)\ell n|x-2| + C$

42. $\dfrac{-10x^2+36x-17}{(x-2)^2(x+1)} = \dfrac{A}{x-2} + \dfrac{B}{(x-2)^2} + \dfrac{C}{x+1} = \dfrac{A(x-2)(x+1) + B(x+1) + C(x-2)^2}{(x-2)^2(x+1)}$

Therefore, $-10x^2+36x-17 = A(x-2)(x+1) + B(x+1) + C(x-2)^2$.

Obtain $B = 5$ by letting $x = 2$; obtain $C = -7$ by letting $x = -1$; obtain $A = -3$ by using the values obtained for B & C and letting $x = 0$.

$\int\left(\dfrac{-3}{x-2} + \dfrac{5}{(x-2)^2} - \dfrac{7}{x+1}\right)dx = -3\ell n|x-2| - 5(x-2)^{-1} - 7\ell n|x+1| + C$

45. $P'(t) = 3te^{-0.4t}$ and $P = 0$ when $t = 0$.

$P(t) = -7.5e^{-0.4t}(t+2.5) + C$ [Use integration by parts with $u = 3t$ & $dv = e^{-0.4t}$]

$P = 0$ when $t = 0$ infers $C = 18.75$, so $P(t) = -7.5e^{-0.4t}(t+2.5) + 18.75$.
Therefore, $P(15) = -7.5e^{-0.4(15)}(17.5) + 18.75 \approx 18.4$ mg.

       Problem Set 7.3

3. $(1-3y^2)dy = x^2dx$, so $\int(1-3y^2)dy = \int x^2dx$.

   $y-y^3 = \frac{1}{3}x^3+C$, or $3y-3y^3-x^3 = K$ (where $K = 3C$).

6. $e^y dy = (1+2x)dx$, so $\int e^y dy = \int(1+2x)dx$.

   $e^y = x+x^2+C$, or $y = \ln(x^2+x+C)$

9. $\frac{1}{y} dy = \frac{2}{x(x-1)} dx = \left[\frac{-2}{x} + \frac{2}{x-1}\right] dx$ (Partial fractions decomposition)

   $\ln|y| = -2\ln|x|+2\ln|x-1|+\ln C = \ln\left[\frac{C(x-1)^2}{x^2}\right]$, $C > 0$

   $y = \frac{C(x-1)^2}{x^2}$, C unrestricted

12. $y^{1/2}dy = x^{1/2}\ell nxdx$, so $\int y^{1/2}dy = \int x^{1/2}\ell nxdx$.

   (On right side, use integration by parts with $u = \ell nx$ & $dv = x^{1/2}dx$.)

   $\frac{2}{3}y^{3/2} = \frac{2}{3}x^{3/2}\ell nx - \frac{4}{9}x^{3/2} + C$, or $3y^{3/2} = 3x^{3/2}\ell nx - 2x^{3/2} + K$.

15. $-y^{-2}dy = t(t^2+2)^4dt$

   $y^{-1} = \frac{1}{2}\int u^4du = \frac{1}{10}u^5+C = \frac{1}{10}(t^2+2)^5+C$ [Letting $u = t^2+2$, $du = 2tdt$]

   or $10y^{-1} = (t^2+2)^5+K$.

18. $y = 10e^{-0.18t}$ (Exponential growth model with $y_0 = 10$ & $k = -0.18$)

21. $y = 12-(12-8)e^{-4t}$ (Bounded growth model with $y_0 = 8$, $L = 12$, $k = 4$)

   $y = 12-4e^{-4t}$

24. $y = \frac{(6)(5)}{5+(6-5)e^{-(6)(0.04)t}}$ (Logistic growth model with $y_0=5$, $L=6$, $k=0.04$)

   $y = \frac{30}{5+e^{-0.24t}}$

27. $y = y_0e^{-4t}$ (Exponential Growth model with $k = -4$)

30. $\frac{1}{y}$ dy = $\frac{\ell nx}{x}$ dx

$\ell n|y| = \int u du = \frac{1}{2}u^2 + C = \frac{1}{2}(\ell nx)^2 + C$ [Letting u = ℓnx, du = $\frac{1}{x}dx$]

$2\ell n|y| = (\ell nx)^2 + K$

33. $\frac{1}{y}$ dy = $\frac{8}{x(x-2)}$ dx = $\left(\frac{-4}{x} + \frac{4}{x-2}\right)$dx   (Partial fractions decomposition)

$\ell n|y| = -4\ell n|x|+4\ell n|x-2|+\ell nC = \ell n\left[\frac{C(x-2)^4}{x^4}\right]$, C > 0

y = $\frac{C(x-2)^4}{x^4}$, C unrestricted

36. Elasticity of demand means that $-\frac{p}{x}\frac{dx}{dp} = 1$, so $-\frac{1}{x}$ dx = $\frac{1}{p}$ dp

$-\ell nx = \ell np + C$, so $\ell nx + \ell np = -C$; $\ell n(xp) = -C$; $xp = e^{-C}$.
That is, the product of x & p is a constant.

39. (a) $\frac{dy}{dt}$ = ky(5000-y)   (Logistic Growth Model with L = 5000, y_0 = 5)

(b) y = $\frac{(5000)(5)}{5 + (5000-5)e^{-5000kt}}$ = $\frac{5000}{1 + 999e^{-5000kt}}$

y = 30 when t = 1. Obtain -5000k = $\ell n(\frac{497}{2997}) \approx$ -1.7968

Therefore, y = $\frac{5000}{1 + 999e^{-1.7968t}}$ (approximately).

If t = 4, y = $\frac{5000}{1 + 999e^{-1.7968(4)}} \approx$ 2850 people.

42. $\frac{dP}{dt}$ = 0.0002(P-500)(2500-P), with P_0 = 1000

$0.0002dt = \frac{1}{(P-500)(2500-P)}$ dP = $\left(\frac{0.0005}{P-500} + \frac{0.0005}{2500-P}\right)$dP

Therefore $0.0002t = 0.0005\ell n(P-500) - 0.0005\ell n(2500-P) + C$

$0.4t = \ell n(P-500) - \ell n(2500-P) + \ell nk = \ell n\left[\frac{k(P-500)}{2500-P}\right]$

t = 0 & P = 1000 infers K = 3. Then let t = 15 and obtain

P = $\frac{1500+2500e^6}{3+e^6} \approx$ 2485 bears.

3. (a) x^3-6x^2+5x+C

   (b) $\int (3x^{1/2}+5x^{-2})dx = 2x^{3/2}-5x^{-1}+C$

   (c) $\int \frac{4x^2+4x+1}{x} dx = \int (4x+4+x^{-1})dx = 2x^2+4x+\ell n|x|+C$

   (d) $\int (4e^{3x}-6x^{-1/3})dx = \frac{4}{3}e^{3x}-9x^{2/3}+C$

6. $F(x) = \frac{1}{4}x^4 + 3x^2 + 12\ell n|x| - 4x^{-2} + 4x + C$

   $4 = F(1) = \frac{1}{4} + 3 + 0 - 4 + 4 + C$, so $C = \frac{3}{4}$

   $\therefore F(x) = \frac{1}{4}x^4 + 3x^2 + 12\ell n|x| - 4x^{-2} + 4x + \frac{3}{4}$

9. $C'(x) = 280-12x+0.6x^2$, so $C(x) = 280x-6x^2+0.2x^3+K$

   $3100 = C(10) = 2800-600+200+K$, so $K = 700$

   $\therefore C(x) = 280x-6x^2+0.2x^3+700$

12. $\frac{dP}{dt} = 150e^{-0.015t}$, with $P_0 = 25000$ [$t = 0$ on 01/01/1990]

   $\therefore P = -10000e^{-0.015t}+C$; then $25000 = -10000+C$, so $C = 35000$

   $\therefore P = -10000e^{-0.015t}+35000$

   $\therefore$ When $t = 10$ (on 01/01/2000), $P = -10000e^{-0.15}+35000 \approx 26393$

   On January 1, 2000, the population will be about 26,400.

15. (a) $\int xe^{4x}dx = \frac{1}{4}xe^{4x} - \int \frac{1}{4}e^{4x}dx$

   $= \frac{1}{4}xe^{4x} - \frac{1}{16}e^{4x} + C$

   | Let $u = x$ & $dv = e^{4x}dx$ |
   | Then $du = dx$ & $v = \frac{1}{4}e^{4x}$ |

   (b) $\int x^{-3/2}\ell nx \, dx = -2x^{-1/2}\ell nx + \int 2x^{-3/2}dx$

   $= -2x^{-1/2}\ell nx - 4x^{-1/2} + C$

   | Let $u = \ell nx$ & $dv = x^{-3/2}dx$ |
   | Then $du = x^{-1}dx$ & $v = -2x^{-1/2}$ |

18. (a) $\dfrac{x-16}{(x+5)(x-2)} = \dfrac{A}{x+5} + \dfrac{B}{x-2} = \dfrac{A(x-2) + B(x+5)}{(x+5)(x-2)}$, so $x-16 = A(x-2)+B(x+5)$.

Therefore, $\dfrac{x-16}{x^2+3x-10} = \dfrac{3}{x+5} + \dfrac{-2}{x-2}$.

| Let x = -5; obtain A = 3. |
| Let x = 2; obtain B = -2. |

$\displaystyle\int\left(\dfrac{3}{x+5} + \dfrac{-2}{x-2}\right)dx = 3\ln|x+5| - 2\ln|x-2| + C$

(b) $\dfrac{x^2-15x+18}{x(x+3)(x-3)} = \dfrac{A}{x} + \dfrac{B}{x+3} + \dfrac{C}{x-3} = \dfrac{A(x+3)(x-3) + Bx(x-3) + Cx(x+3)}{x(x+3)(x-3)}$

Therefore, $x^2-15x+18 = A(x+3)(x-3) + Bx(x-3) + Cx(x+3)$

Obtain A = -2 by letting x = 0; obtain B = 4 by letting x = -3; obtain C = -1 by letting x = 3.

$\displaystyle\int\left(\dfrac{-2}{x} + \dfrac{4}{x+3} + \dfrac{-1}{x-3}\right)dx = -2\ln|x| + 4\ln|x+3| - \ln|x-3| + C$

21. (a) $2y^{-3}dy = e^x dx$; $-y^{-2} = e^x + C$

$y = \frac{1}{3}$ when $x = 0$: $\quad -9 = 1 + C$, so $C = -10$.

$\therefore -y^{-2} = e^x-10$

(b) $-y^{-2}dy = 4x(5+x^2)^{1/2}dx$; $y^{-1} = \frac{4}{3}(5+x^2)^{3/2} + C$

$y = \frac{1}{2}$ when $x = 2$: $\quad 2 = 36 + C$, so $C = -34$

$\therefore y^{-1} = \frac{4}{3}(5+x^2)^{3/2}-34$

24. (a) $\dfrac{dy}{dt} = ky(50000-y)$, $y_0 = 10$

(b) [Logistic growth model with L = 50000, $y_0 = 10$]

$B = \dfrac{50000-10}{10} = 4999$, so $y = \dfrac{50000}{1 + 4999e^{-Lkt}}$

$y = 5000$ when $t = 5$, so $5000 = \dfrac{50000}{1 + 4999e^{-5Lk}}$

Therefore, $Lk = -0.2\ln(\frac{1}{4999}) \approx 1.263953719$

Then, when $t = 12$, $y = \dfrac{50000}{1 + 4999e^{-(1.263953719)(12)}} \approx 49935$.

About 49,900 people had heard the rumor by the end of twelve days.

27. (a) (a) $\frac{dy}{dt} = ky(25000-y)$, $y_0 = 50$

(b) [Logistic growth model with $L = 25000$, $y_0 = 50$]

$B = \frac{25000-50}{50} = 499$, so $y = \frac{25000}{1 + 499e^{-Lkt}}$

$y = 750$ when $t = 2$, so $750 = \frac{25000}{1 + 499e^{-2Lk}}$

Therefore, $Lk = -0.5 \ln(\frac{97}{1497}) \approx 1.368253703$

Then, when $t = 5$, $y = \frac{25000}{1 + 499e^{-(1.368253703)(5)}} \approx 16297$.

About 16,300 people had been infected by the end of five weeks.

Problem Set 8.1   Area and the Definite Integral

3. Hypotenuse has length 5, so circle has radius 2.5.
   Area $= \frac{1}{2}(4)(3) + \frac{1}{2}\pi(2.5)^2 \approx 15.82$ square units.

6. Area $= \pi(6)^2 - \pi(3)^2 \approx 84.82$ square units.

9. The triangle is a right triangle. Its hypotenuse has length $\sqrt{36+64}$
   which is 10, so the radius of the circle is 5. Therefore, the area is
   $\frac{1}{2}\pi(5)^2 - \frac{1}{2}(8)(6) \approx 15.27$ square units.

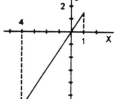

12. $\frac{1}{2}(1)(1.5) - \frac{1}{2}(4)(6) = -11.25$

15. $0 - \frac{1}{2}(2)(6) = -6$

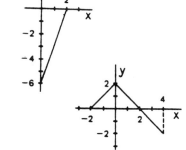

18. $\frac{1}{2}(4)(2) - \frac{1}{2}(2)(2) = 2$. A simpler way is
    to note that the value of the integral of
    $f(x)$ for the interval $[0,4]$ is zero since
    the triangles above and below the x-axis
    are congruent. Then the calculation for
    the integral we want is $\frac{1}{2}(2)(2) + 0 = 2$.

21. $0 + (1)(1) - \frac{1}{2}(2)(2) = -1$.
    [See note with solution of
    Problem 18.]

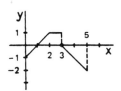

24. $\frac{1}{2}(1)(1.5) - [\frac{1}{2}(2)(3) + \frac{1}{4}\pi(3)^2] = -2.25(1+\pi) \approx -9.32$

27. $\frac{(3)^3}{3} = 9$ 30. $\frac{(\sqrt{2})^4}{4} = 1$

33. (a) Odd since $f(-x) = 2(-x)^3 - (-x) = -2x^3 + x = -(2x^3 - x) = -f(x)$.
    (b) Even since $f(-x) = 2(-x)^2 + 5 = 2x^2 + 5 = f(x)$.
    (c) Neither since, for example, $f(-2) = -1$ & $f(2) = 15$ (not -1 or 1).
    (d) Odd since $f(-x) = \dfrac{(-x)^2 + 4}{(-x)^7 + 4(-x)} = \dfrac{x^2 + 4}{-x^7 - 4x} = -\dfrac{x^2 + 4}{x^7 + 4x} = -f(x)$.

36. $2\int_0^2 x^2 dx = 2[\frac{1}{3}(2)^3] = \frac{16}{3}$ [$f(x) = x^2$ is an even function.]

39. $2\int_0^2 |x| dx = 2\int_0^2 x dx = 2[\frac{1}{2}(2)^2] = 4$ [$f(x) = |x|$ is an even function.]

42. 0   [$f(x) = e^x - e^{-x}$ is an odd function.]

45. $\frac{1}{2}(2)(2) + \frac{1}{2}(1)(1) = 2.5$

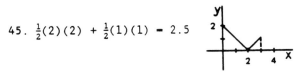

48. $\frac{1}{4}\pi(4)^2 = 4\pi \approx 12.57$

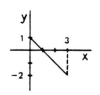

51. $\int_{-2}^3 |t| dx = \frac{1}{2}(2)(2) + \frac{1}{2}(3)(3) = 6.5$

54. $\int_0^4 |5-2t| dt = \frac{1}{2}(2.5)(5) + \frac{1}{2}(1.5)(3) = 8.5$

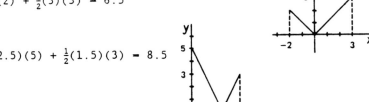

Problem Set 8.2   The Fundamental Theorem of Calculus

3. (i) $\frac{1}{2}(1)(1) - \frac{1}{2}(2)(2) = -1.5$

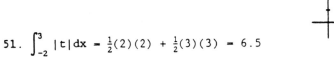

(ii) $\int_0^3 (-x+1) dx = \left[-\frac{1}{2}x^2+x\right]_0^3 = [-\frac{1}{2}(9)+3] - [0+0] = -1.5$

6. $\int_{-2}^3 (-\frac{2}{3}x) dx = \left[-\frac{1}{3}x^2\right]_{-2}^3 = [-3-(-\frac{4}{3})] = -\frac{5}{3}$

9. $\int_1^3 (3x^2+2x-8) dx = \left[x^3+x^2-8x\right]_1^3 = [(27+9-24) - (1+1-8)] = 18$

12. $\int_{-1}^2 (5x^4-3x) dx = \left[x^5-\frac{3}{2}x^2\right]_{-1}^2 = [(32-6) - (-1-\frac{3}{2})] = 28.5$

15. $\int_1^8 4x^{1/3}dx = \left[3x^{4/3}\right]_1^8 = [3(16) - 3(1)] = 45$

18. $\int_1^3 (-2x^{-2})dx = \left[\frac{2}{x}\right]_1^3 = [\frac{2}{3} - 2] = -\frac{4}{3}$

21. Area $= \int_{-1}^3 (x^2+4)dx = \left[\frac{1}{3}x^3+4x\right]_{-1}^3$

$= [(9+12)-(-\frac{1}{3}-4)] \approx 25.333$

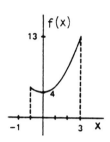

24. Area $= \int_0^4 t^{3/2}dt = \left[\frac{2}{5}t^{5/2}\right]_0^4$

$= \frac{2}{5}[32-0] = 12.8$

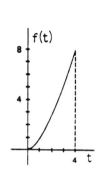

27. $-2\int_{-1}^3 f(x)dx = -2(-8) = 16$

30. $2\int_3^5 g(x)dx - \int_3^5 f(x)dx = 2(2)-(4) = 0$

33. $-2\int_{-1}^5 f(x)dx + \int_{-1}^5 3dx = 2\left[\int_{-1}^3 f(x)dx + \int_3^5 f(x)dx\right] + \left[3x\right]_{-1}^5$

$= 2[-8+4] + [15-(-3)] = 10$

36. $\int_{-1}^7 f(x)dx + \int_7^5 f(x)dx = \int_{-1}^5 f(x)dx = \int_{-1}^3 f(x)dx + \int_3^5 f(x)dx$

$= (-8+4) = -4$

39. $\int_{-2}^0 x^2 dx + \int_0^2 x\,dx + \int_2^3 (4-x^2)dx$

$= \left[\frac{1}{3}x^3\right]_{-2}^0 + \left[\frac{1}{2}x^2\right]_0^2 + \left[4x-\frac{1}{3}x^3\right]_2^3$

$= [0-(-\frac{8}{3})] + [2-0] + [(12-9)-(8-\frac{8}{3})] = \frac{7}{3}$

42. $\dfrac{\int_0^3 (3x^2-4x)\,dx}{3-0} = \dfrac{\left[x^3-2x^2\right]_0^3}{3} = \dfrac{27-18}{3} = 3$

45. $\dfrac{\int_{-1}^2 (4-x^2)\,dx}{2-(-1)} = \dfrac{\left[4x-\frac{1}{3}x^3\right]_{-1}^2}{3} = \dfrac{(8-\frac{8}{3})-(-4+\frac{1}{3})}{3} = 3$

48. $\left[-x^{-1}+2\ln x\right]_1^4 = (-\frac{1}{4}+2\ln 4)-(-1+0) = \frac{3}{4}+2\ln 4 \approx 3.523$

51. $\int_0^2 -2(t-2)\,dt + \int_2^3 2(t-2)\,dt = \left[-t^2+4t\right]_0^2 + \left[t^2-4t\right]_2^3$

$= [(-4+8)-(0-0)] + [(9-12)-(4-8)] = 5$

54. $\left[g(t)\right]_0^4 = g(4)-g(0) = e^2-e^0 = e^2-1 \approx 6.389$

57. $\int 2xe^{x^2}\,dx = \int e^u\,du = e^u+C = e^{x^2}+C$

$\boxed{\begin{array}{l} \text{Let } u = x^2 \\ \text{Then } du = 2x\,dx \end{array}}$

Therefore, $\int_0^1 2xe^{x^2}\,dx = \left[e^{x^2}\right]_0^1 = e^1-e^0 = e-1 \approx 1.718.$

60. $\int xe^x\,dx = xe^x - \int e^x\,dx$

$\boxed{\begin{array}{l} \text{Let } u = x \ \& \ dv = e^x dx \\ \text{Then } du = dx \ \& \ v = e^x \end{array}}$

$= xe^x-e^x + C = e^x(x-1) + C$

Thus, Area $= \int_0^2 xe^x\,dx = \left[e^x(x-1)\right]_0^2 = [e^2(1)-e^0(-1)] = e^2+1 \approx 8.389$

63. $\dfrac{\int_0^{20} (40+50t-0.6t^2)\,dt}{20-0} = \dfrac{\left[40t+25t^2-0.2t^3\right]_0^{20}}{20} = \dfrac{800 + 10000 - 1600}{20}$

$= 460$ pounds

Problem Set 8.3 Calculation of Definite Integrals

3. (i) $\int_0^2 (x^5 - 6x^3 + 9x)\,dx = \left[\frac{1}{6}x^6 - \frac{3}{2}x^4 + \frac{9}{2}x^2\right]_0^2 = [(\frac{32}{3} - 24 + 18) - (0)] = \frac{14}{3}$

   (ii) $\int_0^2 (x^2 - 3)^2 x\,dx = \frac{1}{2}\int_{-3}^1 u^2\,du$

   $\qquad = \frac{1}{2}\left[\frac{1}{3}u^3\right]_{-3}^1 = \frac{1}{6}[(1) - (-27)] = \frac{14}{3}$

   > Let $u = x^2 - 3$
   > Then $du = 2x\,dx$
   > $x=2 \Rightarrow u=1$
   > $x=0 \Rightarrow u=-3$

6. $\left[\frac{1}{12}(3x-2)^4\right]_0^1 = \frac{1}{12}[(1) - (16)] = -1.25$ [Could use $u = 3x-2$.]

9. $\left[5\ln(x-4)\right]_5^8 = 5[\ln 4 - \ln 1] = 5\ln 4 \approx 6.931$ [Could use $u = x-4$.]

12. $\left[\frac{1}{2}\ln(x^2-8)\right]_3^5 = \frac{1}{2}[\ln 17 - 0] = \frac{1}{2}\ln 17 \approx 1.417$ [Could use $u = x^2-8$.]

15. $\left[-\frac{1}{2}e^{-x^2}\right]_0^2 = -\frac{1}{2}[e^{-4} - 1] \approx 0.4908$ [Could use $u = e^{-x^2}$.]

18. $\left[\frac{1}{15}(x^3-6x)^5\right]_0^2 = \frac{1}{15}[(-4)^5 - (0)] \approx -68.267$ [Could use $u = x^3-4$.]

21. $\left[\frac{1}{9}[\ln(1+3x)]^3\right]_0^3 = \frac{1}{9}[(\ln 10)^3 - (0)] \approx 1.356$ [Could use $u = \ln(1+3x)$.]

24. $\left[\frac{1}{4}(x+2e^{x/2})^4\right]_0^2 = \frac{1}{4}[(2+2e)^4 - (2)^4] \approx 760.591$ [Could use $u = x+2e^{x/2}$.]

27. $\left[\frac{2}{5}x^{5/2}\ln x\right]_1^e - \int_1^e \frac{2}{5}x^{3/2}\,dx$

   > Letting $u = \ln x$ & $dv = x^{3/2}dx$
   > Then $du = x^{-1}dx$ & $v = \frac{2}{5}x^{5/2}$

   $\qquad = [\frac{2}{5}e^{5/2} - 0] - \left[\frac{4}{25}x^{5/2}\right]_1^e = \frac{2}{5}e^{5/2} - [\frac{4}{25}e^{5/2} - \frac{4}{25}] = \frac{6}{25}e^{5/2} + \frac{4}{25} \approx 3.084$

30. $\left[\frac{1}{7}x(x-2)^7\right]_0^2 - \int_0^2 \frac{1}{7}(x-2)^7\,dx$

   > Letting $u = x$ & $dv = (x-2)^6dx$
   > Then $du = dx$ & $v = \frac{1}{7}(x-2)^7$

   $\qquad = [0-0] - \left[\frac{1}{56}(x-2)^8\right]_0^2 = -\frac{1}{56}[0 - 256] = \frac{32}{7} \approx 4.571$

33. $\left[\frac{x}{2}\sqrt{x^2+16} - \frac{16}{2}\ln(x+\sqrt{x^2+16})\right]_0^3$ [Formula 49 with u - x & a - 4]

     $- [(\frac{15}{2}-8\ln8) - (0-8\ln4)] - \frac{15}{2}-8\ln2 \approx 1.955$

36. $\left[\frac{2}{15(9)}(9x-2)(3x+1)^{3/2}\right]_0^1$ [Formula 96 with u - x, a - 3, & b - 1]

     $- \frac{2}{135}[(7)(8) - (-2)(1)] - \frac{116}{135} \approx 0.859$

39. $\left[\frac{u}{2}\sqrt{u^2-9} + \frac{9}{2}\ln(u+\sqrt{u^2-9})\right]_5^{\sqrt{34}}$ [Formula 49 with a - 3]

     $- \frac{5}{2}\sqrt{34} - 10 + \frac{9}{2}\ln[\frac{1}{9}(5+\sqrt{34})] \approx 5.411$

42. $\int_0^2 (x+1)^2 dx - \left[\frac{1}{3}(x+1)^3\right]_0^2 - \frac{1}{3}[27-1] - \frac{26}{3} \approx 8.667$

45. $\dfrac{\int_1^{\sqrt[3]{2}}(x^3-1)^4 x^2 dx}{\sqrt[3]{2}-1} - \dfrac{\left[\frac{1}{15}(x^3-1)^5\right]_1^{\sqrt[3]{2}}}{\sqrt[3]{2}-1} - \dfrac{\frac{1}{15}[1-0]}{\sqrt[3]{2}-1} \approx 0.2565$

Problem Set 8.4   The Definite Integral as a Limit

3. $5 + 8 + 11 + 14 - 38$ 6. $\frac{3}{2} + \frac{4}{5} + \frac{1}{2} - 2.8$

9. $\sum_{i=1}^{10} i^3$ 12. $\sum_{i=1}^{70} \frac{5}{i}$

15. $\sum_{i=1}^{n} g(t_i)\Delta t$ 18. $\sum_{i=1}^{n} x_i^2 \Delta x$

21. $4\sum_{i=1}^{10} x_i = 4(30) = 120$ 24. $2\sum_{i=1}^{10} x_i + 11\sum_{i=1}^{10} y_i = 2(30) + 11(4) = 104$

27. $2\sum_{i=1}^{10} x_i - 3\sum_{i=1}^{10} y_i - 3\sum_{i=1}^{10} 1 = 2(30) - 3(4) - 3(10) = 18$

30. $\displaystyle\lim_{n\to\infty}\left[\sum_{i=1}^{n}f(x_i)\Delta x\right] = \lim_{n\to\infty}\left[\sum_{i=1}^{n}(6 - \frac{4}{n}i)\frac{4}{n}\right] = \lim_{n\to\infty}\left[\frac{24}{n}\sum_{i=1}^{n}1 - \frac{16}{n^2}\sum_{i=1}^{n}i\right]$

$= \lim_{n\to\infty}\left[\frac{24}{n}(n) - \frac{16}{n^2}\frac{n(n+1)}{2}\right] = \lim_{n\to\infty}[24 - 8(1+\frac{1}{n})] = 24-8 = 16$

33. $\displaystyle\lim_{n\to\infty}\left[\sum_{i=1}^{n}f(x_i)\Delta x\right] = \lim_{n\to\infty}\left[\sum_{i=1}^{n}(\frac{9}{n^2}i^2+1)\frac{3}{n}\right] = \lim_{n\to\infty}\left[\frac{27}{n^3}\sum_{i=1}^{n}i^2 - \frac{3}{n}\sum_{i=1}^{n}1\right]$

$= \lim_{n\to\infty}\left[\frac{27}{n^3}\frac{n(n+1)(2n+1)}{6} + \frac{3}{n}(n)\right] = \lim_{n\to\infty}[\frac{9}{2}(1+\frac{1}{n})(2+\frac{1}{n}) + 3] = 9+3 = 12$

36. $\displaystyle\lim_{n\to\infty}\left[\sum_{i=1}^{n}f(x_i)\Delta x\right] = \lim_{n\to\infty}\left[\sum_{i=1}^{n}(\frac{2}{n^2}i^2 + \frac{1}{n}i)\frac{1}{n}\right] = \lim_{n\to\infty}\left[\frac{2}{n^3}\sum_{i=1}^{n}i^2 + \frac{1}{n^2}\sum_{i=1}^{n}i\right]$

$= \lim_{n\to\infty}\left[\frac{2}{n^3}\frac{n(n+1)(2n+1)}{6} + \frac{1}{n^2}\frac{n(n+1)}{2}\right] = \lim_{n\to\infty}[\frac{1}{3}(1+\frac{1}{n})(2+\frac{1}{n}) + \frac{1}{2}(1+\frac{1}{n})] = \frac{2}{3} + \frac{1}{2} = \frac{7}{6}$

39. $\displaystyle\sum_{i=1}^{499}i = \frac{(499)(500)}{2} = 124{,}750$

42. $\displaystyle\sum_{i=1}^{100}i^2 - \sum_{i=1}^{40}i^2 = \frac{(100)(101)(201)}{6} - \frac{(40)(41)(81)}{6} = 316{,}210$

45. $\displaystyle 2\sum_{k=1}^{100}1 - 3\sum_{k=1}^{100}k + \sum_{k=1}^{100}k^2 = 2(100) - 3\frac{(100)(101)}{2} + \frac{(100)(101)(201)}{6} = 323{,}400$

48. $\displaystyle\int_{0}^{2}(3x^2+x-2)\,dx$ [where $\Delta x = \frac{2}{n}$ and $x_i = \frac{2}{n}i$]

51. $\bar{x} = \frac{1}{10}(3 + 4 + 5 + 5 + 6 + 6 + 6 + 7 + 8 + 10) = 6$

$s^2 = \frac{1}{10}[(3-6)^2 + (4-6)^2 + (5-6)^2 + (5-6)^2 + (6-6)^2 + (6-6)^2 + (6-6)^2 +$

$+ (7-6)^2 + (8-6)^2 + (10-6)^2]$

$= \frac{1}{10}[9 + 4 + 1 + 1 + 0 + 0 + 0 + 1 + 4 + 16] = 3.6$

3. (a) $\frac{1}{2}(3)(3) - \frac{1}{2}(2)(2) = 2.5$

   (b) $\frac{1}{2}\pi(2)^2 = 2\pi \approx 6.2832$

   (c) $[\frac{1}{2}(2)(2) + \frac{1}{2}(3)(2)] - [\frac{1}{2}(4+3)(2)] = -2$

   (a)   (b)   (c)

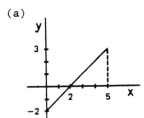 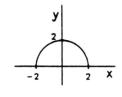 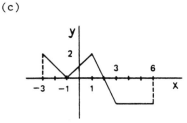

6. (a) 0, since f is an odd function.
   (b) 2(3) = 6, since |f(x)| defines an even function.
   (c) 2(3) = 6, since g is an even function.
   (d) 0, since the product of an odd function and an even function is an
       odd function.

9. $\int_1^9 3x^{1/2}dx = \left[2x^{3/2}\right]_1^9 = [54-2] = 52$

12. $\int_0^2 (3x^2-\frac{1}{2}x+4)dx = \left[x^3-\frac{1}{4}x^2+4x\right]_0^2 = [(8-1+8)-(0)] = 15$

15. $\left[-e^{1/x}\right]_{1/4}^{1/2} = -e^2+e^4 = e^4-e^2 \approx 47.2091$

    [Could use u-substitution with $u = 1/x$.]

18. $\int_2^5 (2x^2+x-6)^{-3/2}(4x+1)dx = \left[-2(2x^2+x-6)^{-1/2}\right]_2^5 = -\frac{2}{7} + 1 = \frac{5}{7}$

    [Could use u-substitution with $u = (2x^2+x-6)$.]

21. $\int_0^3 -4(t-3)dt + \int_3^5 4(t-3)dt = -4\left[\frac{1}{2}t^2-3t\right]_0^3 + 4\left[\frac{1}{2}t^2-3t\right]_3^5$

    $= -4[-\frac{9}{2} - 0] + 4[-\frac{5}{2} + \frac{9}{2}] = 26$

24. $\left[\left[\ell n(2x+1)\right]^2\right]_0^{(e-1)/2} = 1 - 0 = 1$

[Could use u-substitution with $u = \ell n(2x+1)$.]

27. $\displaystyle\int_{-1}^{0} u^5(u+3)\,du = \int_{-1}^{0} (u^6 + 3u^5)\,du$

$\qquad = \left[\tfrac{1}{7}u^7 + \tfrac{1}{2}u^6\right]_{-1}^{0} = [0 - (-\tfrac{1}{7} + \tfrac{1}{2})] = \tfrac{-5}{14}$

> Let $u = x-3$, so $x = u+3$
> Then $du = dx$
> $x=3 \Rightarrow u=0$
> $x=2 \Rightarrow u=-1$

30. $\left[\tfrac{x}{8}(2x^2+45)\sqrt{x^2+9} + \tfrac{243}{8}\ell n|x+\sqrt{x^2+9}|\right]_0^4 = 192.5 + \tfrac{243}{8}\ell n9 - \tfrac{243}{8}\ell n3 \approx 225.8703$

[Using Formula 53]

33. $\displaystyle\int_0^{30} (52 + 3.2t - 0.02t^2)\,dt = \left[52t + 1.6t^2 - \tfrac{0.02}{3}t^3\right]_0^{30} = 2820$

Therefore, $h_{ave} = \dfrac{2820}{30-0} = 94$ inches.

36. $\displaystyle\sum_{k=0}^{4} k^3 - \sum_{k=0}^{4} 2^k = (0 + 1 + 8 + 27 + 64) - (1 + 2 + 4 + 8 + 16) = 69$

39. $4\displaystyle\sum_{i=1}^{8} x_i - 6\sum_{i=1}^{8} y_i = 4(36) - 6(15) = 54$

42. $\displaystyle\sum_{i=1}^{10} i^3 - 2\sum_{i=1}^{10} i^2 - 3\sum_{i=1}^{10} i = (55)^2 - 2(385) - 3(55) = 2090$

45. $\displaystyle\lim_{n\to\infty}\left[\sum_{i=1}^{n}\left(\tfrac{8}{n}i + 3\right)\tfrac{2}{n}\right]$ $\qquad [\Delta x = \tfrac{2}{n},\ x_i = \tfrac{2}{n}i]$

$\qquad = \displaystyle\lim_{n\to\infty}\left[\tfrac{16}{n^2}\sum_{i=1}^{n} i + \tfrac{6}{n}\sum_{i=1}^{n} 1\right] = \lim_{n\to\infty}\left[\tfrac{16}{n^2}\tfrac{n(n+1)}{2} + \tfrac{6}{n}(n)\right] = \lim_{n\to\infty}\left[8\left(1 + \tfrac{1}{n}\right) + 6\right]$

$\qquad = 8(1) + 6 = 14$

45. (continued)

Two methods of checking: (1) Use the fundamental theorem of calculus,
(2) Calculate the area of the trapezoidal
region between the graphs of $y = 4x+3$,
the x-axis, and the lines $x=0$ and $x=2$.

Each yields the number 14.

48. It is $\int_0^2 (3x^2+x)\,dx = \left[x^3 + \frac{1}{2}x^2\right]_0^2 = (8+2) - (0) = 10$

Problem Set 9.1   Areas Between Two Curves

3. $\left[\frac{1}{3}(x+1)^3\right]_1^3 = \frac{1}{3}[64-8] = \frac{56}{3} \approx 18.667$

6. $\left[\frac{9}{2}x^2 - \frac{1}{4}x^4\right]_1^3 = \frac{9}{2}[9-1] - \frac{1}{4}[81-1] = 16$

9. $\left[\ln(x+3)\right]_{-2}^0 = [\ln 3 - 0] = \ln 3 \approx 1.099$

12. $\left[-\frac{1}{3}(4-x^2)^{3/2}\right]_0^2 = -\frac{1}{3}[0-8] = \frac{8}{3} \approx 2.667$

15. $\displaystyle\int_0^2 [(x^2+2)-(-2x^2)]dx$

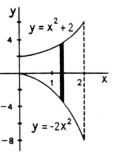

$= \displaystyle\int_0^2 (3x^2+2)dx = \left[x^3+2x\right]_0^2$

$= [(8+4)-(0+0)] = 12$

18. $\displaystyle\int_{-2}^1 [(4-x^2)-(2x^3)]dx$

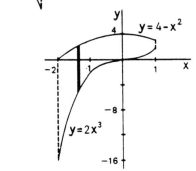

$= \displaystyle\int_{-2}^1 (-2x^3-x^2+4)dx = \left[-\frac{1}{2}x^4 - \frac{1}{3}x^3+4x\right]_{-2}^1$

$= [(-\frac{1}{2}-\frac{1}{3}+4) - (-8+\frac{8}{3}-8)] = 16.5$

21. $\displaystyle\int_{-2}^2 [6(x+3)^{-1}-(-x+1)]dx$

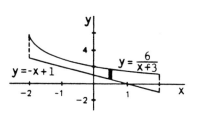

$= \displaystyle\int_{-2}^2 [6(x+3)^{-1} -1]dx = \left[y\ln(x+3)-x\right]_{-2}^2$

$= [(6\ln 5 - 2) - (0+2)]$
$= 6\ln 5 - 4 \approx 5.657$

[Note that the x term was omitted in
the second line, since the integral
of an odd function on -2 to 2 is 0.]

24. Points of intersection are $(-1,0)$ & $(3,8)$.

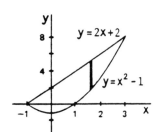

$$\int_{-1}^{3} [(2x+2)-(x^2-1)]\,dx$$

$$= \int_{-1}^{3} (-x^2+2x+3)\,dx = \left[-\tfrac{1}{3}x^3+x^2+3x\right]_{-1}^{3}$$

$$= [(-9+9+9) - (\tfrac{1}{3}+1-3)] = \tfrac{32}{3}$$

27. Points of intersection are $(1,0)$ & $(3,2)$.

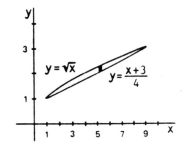

$$\int_{1}^{3} [(x-1)-(x^2-3x+2)]\,dx$$

$$= \int_{1}^{3} (-x^2+4x-3)\,dx = \left[-\tfrac{1}{3}x^3+2x^2-3x\right]_{1}^{3}$$

$$= [(-9+18-9) - (-\tfrac{1}{3}+2-3)] = \tfrac{4}{3}$$

30. Points of intersection are $(1,1)$ & $(9,3)$.

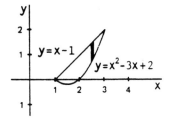

$$\int_{1}^{9} [x^{1/2}-\tfrac{1}{4}(x+3)]\,dx = \left[\tfrac{2}{3}x^{3/2}-\tfrac{1}{8}(x+3)^2\right]_{1}^{9}$$

$$= [(18-18) - (\tfrac{2}{3}-2)] = \tfrac{4}{3}$$

33. Points of intersection are $(-2,8)$, $(0,0)$, & $(2,8)$. Make use of symmetry, which results in the congruence of the regions above and below the x-axis.

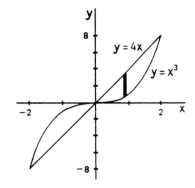

$$2\int_{0}^{2} (4x-x^3)\,dx$$

$$= 2\left[2x^2-\tfrac{1}{4}x^4\right]_{0}^{2}$$

$$= 2[(8-4) - (0-0)] = 8$$

36. Points of intersection are $(-1,1)$, $(0,0)$, & $(2,4)$.

$$\int_{-1}^{0}[(x^3-2x)-x^2]dx + \int_{0}^{2}[x^2-(x^3-2x)]dx$$

$$= \int_{-1}^{0}(x^3-x^2-2x)dx + \int_{0}^{2}(-x^3+x^2+2x)dx$$

$$= \left[\tfrac{1}{4}x^4-\tfrac{1}{3}x^3-x^2\right]_{-1}^{0} + \left[-\tfrac{1}{4}x^4+\tfrac{1}{3}x^3+x^2\right]_{0}^{2}$$

$$= \tfrac{37}{12} \approx 3.083$$

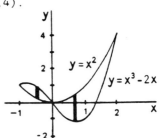

39. Points of intersection are $(0,0)$ & $(3,3)$.

$$\int_{0}^{3}[(4y-y^2)-y]dy - \int_{0}^{3}(-y^2+3y)dy$$

$$= \left[-\tfrac{1}{3}y^3+\tfrac{3}{2}y^2\right]_{0}^{3}$$

$$= [(-9+\tfrac{27}{2}) - 0] = 4.5$$

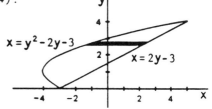

42. Points of intersection are $(-3,0)$ & $(5,4)$.

$$\int_{0}^{4}[(2y-3)-(y^2-2y-3)]dy$$

$$= \int_{0}^{4}(-y^2+4y)dy = \left[-\tfrac{1}{3}y^3+2y^2\right]_{0}^{4}$$

$$= [(-\tfrac{64}{3}+32) - 0] = \tfrac{32}{3}$$

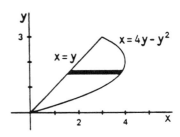

45. Points of intersection are $(-2,0)$, $(0,0)$, & $(3,0)$.

$$\int_{-2}^{0}(x^3-x^2-6x)dx - \int_{0}^{3}(x^3-x^2-6x)dx$$

$$= \left[\tfrac{1}{4}x^4-\tfrac{1}{3}x^3-3x^2\right]_{-2}^{0} - \left[\tfrac{1}{4}x^4-\tfrac{1}{3}x^3-3x^2\right]_{0}^{3}$$

$$= [0-(4+\tfrac{8}{3}-12)] - [(\tfrac{81}{4}-9-27)-0]$$

$$= \tfrac{253}{12} \approx 21.083$$

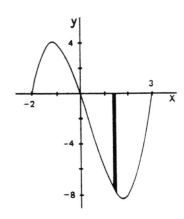

48. Points of intersection are $(0,0)$, $(3,3)$, & $(4,2)$.

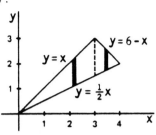

$$\int_0^3 (x - \tfrac{1}{2}x)\,dx + \int_3^4 [(6-x) - \tfrac{1}{2}x]\,dx$$

$$= \int_0^3 \tfrac{1}{2}x\,dx + \int_3^4 (6 - \tfrac{3}{2}x)\,dx$$

$$= \left[\tfrac{1}{4}x^2\right]_0^3 + \left[6x - \tfrac{3}{4}x^2\right]_3^4$$

$$= [\tfrac{9}{4} - 0] + (24 - 12) - (18 - \tfrac{27}{4})] = 3$$

51. Points of intersection are $(-1, -\tfrac{1}{2})$, $(0,0)$, & $(1, \tfrac{1}{2})$.
Make use of congruence.

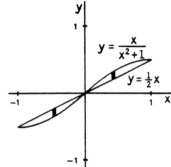

$$2\int_0^1 [x(x^2+1)^{-1} - \tfrac{1}{2}x]\,dx$$

$$= 2\left[\tfrac{1}{2}\ln(x^2+1) - \tfrac{1}{4}x^2\right]_0^1$$

$$= 2[(\tfrac{1}{2}\ln 2 - \tfrac{1}{4}) - 0] = \ln 2 - \tfrac{1}{2} \approx 0.1931$$

54. Make use of congruence.

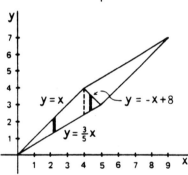

$$2\int_0^4 (x - \tfrac{3}{5}x)\,dx + 2\int_4^5 [(-x+8) - (\tfrac{3}{5}x)]\,dx$$

$$= 2\int_0^4 \tfrac{2}{5}x\,dx + 2\int_4^5 (-\tfrac{8}{5}x+8)\,dx$$

$$= 2\left[\tfrac{1}{5}x^2\right]_0^4 + 2\left[-\tfrac{4}{5}x^2 + 8x\right]_4^5$$

$$= 2[\tfrac{16}{5} - 0] + 2[(-20+40) - (-\tfrac{64}{5} + 32)] = 8$$

Problem Set 9.2   Volumes of Solids of Revolution

3. $\displaystyle\int_0^3 \pi(6x - x^2)^2\,dx = \pi\int_0^3 (x^4 - 12x^3 + 36x^2)\,dx$

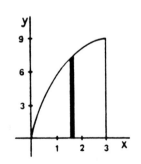

$$= \pi\left[\tfrac{1}{5}x^5 - 3x^4 + 12x^3\right]_0^3$$

$$= \pi[(\tfrac{243}{5} - 243 + 324) - 0]$$

$$= 129.6\pi \approx 407.15$$

6. $\int_0^3 \pi(2-\frac{2}{3}x)^2 dx = \pi\int_0^3(\frac{4}{9}x^2-\frac{8}{3}x+4)dx$

$= \pi\left[\frac{4}{27}x^3-\frac{4}{3}x^2+4x\right]_0^3$

$= \pi[12-12+4] = 4\pi \approx 12.566$

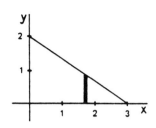

9. $\int_1^4 \pi(\frac{4}{x})^2 dx = 16\pi\int_1^4 x^{-2} dx$

$= 16\pi\left[-\frac{1}{x}\right]_1^4 = 16\pi[-\frac{1}{4}+1]$

$= 12\pi \approx 37.699$

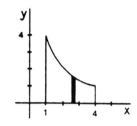

12. $\int_{-2}^0 \pi(e^{-x})^2 dx = \pi\int_{-2}^0 e^{-2x}dx$

$= \pi\left[-\frac{1}{2}e^{-2x}\right]_{-2}^0 = -\frac{\pi}{2}[1-e^4] \approx 84.192$

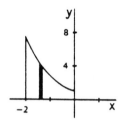

15. $\int_0^1 \pi[(x)^2-(x^2)^2]dx = \pi\int_0^1(x^2-x^4)dx$

$= \pi\left[\frac{1}{3}x^3-\frac{1}{5}x^5\right]_0^1 = \pi[(\frac{1}{3}-\frac{1}{5})-0]$

$= \frac{2\pi}{15} \approx 0.4189$

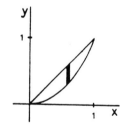

18. Points of intersection are
$(-2,-8)$, $(0,0)$, & $(2,8)$.
Make use of congruence.

$2\int_0^2 \pi[(4x)^2-(x^3)^2]dx = 2\pi\int_0^2(16x^2-x^6)dx$

$= 2\pi\left[\frac{16}{3}x^3-\frac{1}{7}x^7\right]_0^2 = \frac{1024\pi}{15} \approx 153.190$

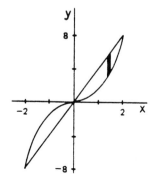

21. Points of intersection are
    $(0,0)$, $(0,6)$, & $(3,0)$.

    $\int_0^6 \pi(-\frac{1}{2}y+3)^2\,dy \; - \; \pi\int_0^6 (\frac{1}{4}y^2-3y+9)\,dy$

    $= \pi\left[\frac{1}{12}y^3-\frac{3}{2}y^2+9y\right]_0^6$

    $= \pi[(18-54+54)-0] = 18\pi \approx 56.549$

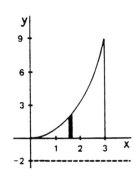

24. $\int_0^9 \pi[(3)^2-(\sqrt{y})^2]\,dy = \pi\int_0^9 (9-y)\,dy$

    $= \pi\left[9y-\frac{1}{2}y^2\right]_0^9 = \pi[81-40.5]$

    $= 40.5\pi \approx 127.235$

27. $\int_0^3 \pi[(x^2+2)^2-(2)^2]\,dx = \pi\int_0^3 (x^4+4x^2)\,dx$

    $= \pi\left[\frac{1}{5}x^5+\frac{4}{3}x^3\right]_0^3 = \pi[(48.6+36)-0]$

    $= 84.6\pi \approx 265.779$

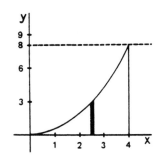

30. $\int_0^4 \pi[(8)^2-(8-x^{3/2})^2]\,dx = \pi\int_0^4 (16x^{3/2}-x^3)\,dx$

    $= \pi\left[\frac{32}{5}x^{5/2}+\frac{1}{4}x^4\right]_0^3 = \pi[(204.8-64)-0]$

    $= 140.8\pi \approx 442.336$

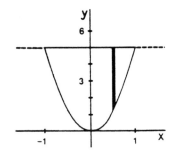

33. $2\int_0^1 \pi(5-5x^2)^2\,dx = 2\pi\int_0^1 (25x^4-50x^2+25)\,dx$

    $= 2\pi\left[5x^5-\frac{50}{3}x^3+25x\right]_0^1$

    $= 2\pi[(5-\frac{50}{3}+25)-0] = \frac{80\pi}{3} \approx 83.776$

36. $\int_0^4 \pi \left([3-(-\frac{1}{2}y+2)]^2 - [3-(\sqrt{4-y})]^2 \right) dy$

$= \pi \int_0^4 [\frac{1}{4}y^2 + 2y + 6(4-y)^{1/2} - 12] dy$

$= \pi \left[\frac{1}{12}y^3 + y^2 - 4(4-y)^{3/2} - 12y \right]_0^4$

$= \pi [(\frac{16}{3} + 16 - 0 - 48) - (0+0-32-0)]$

$= \frac{16\pi}{3} \approx 16.755$

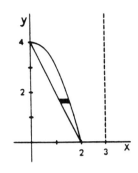

39. $\int_0^{e^4} \pi (\sqrt{\ell ny})^2 dy = \pi \int_0^{e^4} (\ell ny) dy$

$= \pi \left[y(\ell ny - 1) \right]_0^{e^4} = \pi [e^4(4-1) - 0]$

$= 3\pi e^4 \approx 514.575$

[Could use integration by parts with
u = ℓny & dv = dy to obtain the antiderivative given in the second line.]

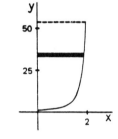

Problem Set 9.3   Social Science Applications

3. $\int_0^8 (450 + 35x^{4/3}) dx = 5520$, so the change is about 5520. The population
will then be about 55,520.

6. (a) $C(180) = 8000 + \int_0^{180} 300e^{-x/90} dx \approx \$31,346$

   (b) Average cost is about $\frac{31346}{180} \approx \$174.14$

9. $P'(x) = R'(x) - C'(x) = (50 + 0.36x - 0.003x^2) - (30 - 0.012x)$
   $= -0.003x^2 + 0.372x + 20$ which equals 0 if $x \approx 165$ units.
   (Use the quadratic formula to obtain that result.)

12. $9 - 0.25x = 0.5x$ if $x = 12$, so equilibrium price is $0.5(12) = \$6$.

15. $240 - 30x - 20x^2 = 20x^2 + 10x$ if $x = 2$, so equilibrium price is
    $20(2)^2 + 10(2) = \$100$.

18. (See Problem 12.)

CS $= \frac{1}{2}(3)(12) = \$18$

PS $= \frac{1}{2}(6)(12) = \$36$

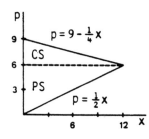

21. (See Problem 15.)

CS $= \int_0^2 [(240-30x-20x^2)-100]dx$

$\approx \$167$

PS $= \int_0^2 [100-(20x^2+10x)]dx$

$\approx \$127$

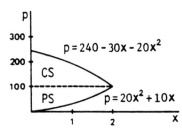

24. Let $45e^{-x/30} = 15$. Then $x = 30\ln 3$

Therefore, CS $= \int_0^{30\ln 3}(45e^{-x/30}-15)dx = 900-450\ln 3 \approx \$406$.

27. $g = 2\int_0^1 (x-x^3)dx = 0.5$

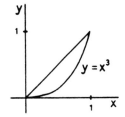

30. $g = 2\int_0^1 [x-(\frac{2}{3}x^2+\frac{1}{3}x)]dx = \frac{2}{9}$

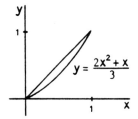

33. $\int_{64000}^{125000}(0.12+0.01x^{1/3})dx = \$34,995$

36. $\int_4^5 100t(t^2+9)^{1/2}dt \approx 2,442$ tons   [Could use $u = t^2+9$]

3. PV = $2400e^{-0.091(2)} + 6200e^{-0.091(5)}$ = \$5,934.22

6. PV(Karen) = $34000e^{-0.085(8)}$ = \$17,224.98

   PV(Robert) = $40000e^{-0.085(10)}$ = \$17,096.60

   Therefore, Karen owes \$128.38 more than Robert.

9. PV = $\int_0^{10} (25000+350t)e^{-0.095t}dt$

   $\hspace{1.5em} = 25000\int_0^{10}e^{-0.095t}dt + 350\int_0^{10} te^{-0.095t}dt \hspace{2em}$ [Use $u = t$ & $dv = e^{-0.095t}dt$
   $\hspace{25em}$ for the second integral.]

   $\hspace{1.5em} = \dfrac{109000 - 122300e^{-0.95}}{0.361}$ = \$170,918.48

12. $\int_1^{\infty}3t^{-4}\,dt = \lim_{b\to\infty}\int_1^{b}3t^{-4}\,dt = \lim_{b\to\infty}\left[-t^{-3}\right]_1^{b} = \lim_{b\to\infty}\left[-b^{-3}-(-1)\right] = 0+1 = 1$

15. $\lim_{b\to\infty}\int_0^{b}te^{-0.09t}\,dt = \lim_{b\to\infty}\left[-\tfrac{100}{9}e^{-0.09t}(t+\tfrac{100}{9})\right]_0^{b} = \lim_{b\to\infty}\left[-\tfrac{100}{9}e^{-0.09b}(b+\tfrac{100}{9}) + \tfrac{10000}{81}\right]$

   $\hspace{1.5em} = \tfrac{10000}{81} \approx 123.457$ [Use $u = t$ & $dv = e^{-0.09t}dt$ for the integration.]

18. PV = $\int_0^{\infty}1200e^{-0.075t}dt = 1200(\tfrac{40}{3})$ = \$16,000

21. PV = $16500e^{-0.10(6)}$ = \$9,055.39

24. $\int_0^{10} (500+20t)e^{-0.08t}dt = 9375-11875e^{-0.8}$ = \$4,039.22

27. The integral diverges since $e^{0.03x} \to \infty$ as $x\to\infty$.

30. $\lim\limits_{b\to\infty}\int_4^b 4(x-3)^{-5}\,dx = \lim\limits_{b\to\infty}\left[-(x-3)^{-4}\right]_4^b = \lim\limits_{b\to\infty}\left[-(b-3)^{-4}+1\right] = 0+1 = 1.$

It converges to 1.

33. $\lim\limits_{b\to\infty}\int_1^b x^{-P}\,dx = \lim\limits_{b\to\infty}\left[\frac{1}{1-p}x^{1-p}\right]_1^b = \lim\limits_{b\to\infty}\left[\frac{1}{1-p}b^{1-p} - \frac{1}{1-p}\right] = 0 - \frac{1}{1-p} = \frac{1}{p-1}$

Problem Set 9.5 Continuous Probability

3. Let **X** be the waiting time. It is uniformly distributed for $0 \le X \le 30$ so $f(x) = \frac{1}{30}$. Prob$(20 \le X \le 30) = \frac{1}{30}(30-20) = \frac{1}{3}$.

6. (a) $1 = \int_1^4 kx^{1/2}dx = \frac{14k}{3}$, so $k = \frac{3}{14}$

   (b) Prob$(2 \le X \le 3) = \int_2^3 \frac{3}{14}x^{1/2}dx = \frac{1}{7}(3^{3/2}-2^{3/2}) \approx 0.3382$

9. (a) $\int_0^6 \frac{36-x^2}{144}\,dx = \frac{1}{144}\left[36x-\frac{1}{3}x^3\right]_0^6 = \frac{1}{144}[216-72] = 1$ (b)

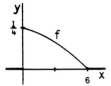

   (c) $\int_1^6 \frac{36-x^2}{144}\,dx = \frac{1}{144}\left[36x-\frac{1}{3}x^3\right]_1^6 = \frac{1}{144}[(216-72) - (36-\frac{1}{3}] = \frac{325}{432} \approx 0.7523$

   (d) $\mu = \int_0^6 xf(x)\,dx = \int_0^6 \frac{36x-x^3}{144}\,dx = \frac{1}{144}\left[18x^2-\frac{1}{4}x^4\right]_0^6 = \frac{1}{144}[648-324] = 2.25$

12. (a) $1 = \int_0^6 2ke^{-kx}dx = \left[-2e^{-kx}\right]_0^6 = -2e^{-6k}+2$, so $k = \frac{\ln 2}{6}$

   (b) $\int_4^6 \frac{\ln 2}{6}e^{(-\frac{1}{6}\ln 2)x}\,dx = -2[2^{-1}-2^{-2/3}] = 2^{1/3}-1 \approx 0.2599$

15. (a) $1 = \int_0^2 k(2x-x^2)\,dx = \frac{4k}{3}$, so $k = \frac{3}{4}$.

15. (b) $\int_0^{0.3} \frac{3}{4}(2x-x^2)dx = \frac{3}{4}[0.09 - 0.009] = 0.06095$

(c) $\mu = \int_0^2 xf(x)dx = \int_0^2 \frac{3}{4}(2x^2-x^3)dx = \frac{3}{4}[\frac{16}{3}-4] = 1$

18. $f(x) = \frac{1}{b}$, so $\frac{1}{5} = \text{prob}(X > 4) = \int_4^b \frac{1}{b}dx = 1-\frac{4}{5}$, so $b = 5$.

21. $\int_0^\infty ke^{-kx}dx = \lim_{b \to \infty}\left[-e^{-kx}\right]_0^b = \lim_{b \to \infty}[-e^{-kx}+1] = 0+1 = 1$, for all $k > 0$.

$\mu = \int_0^\infty xf(x)dx = \lim_{b \to \infty}\int_0^b kxe^{-kx}dx = \frac{1}{k}$ [Use $u = x$ & $dv = ke^{-kx}dx$]

Chapter 9 Review Problem Set

3. $\int_0^2 [(x^3+1)-(2x-3)]dx$

$= \int_0^2 (x^3-2x+4)dx = 8$

6. $\int_1^{\ln 4} (4-e^x)dx = e - 8 + 4\ln 4 \approx 0.2635$

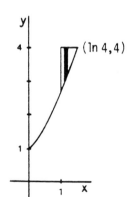

$(\ln 4, 4)$

9. (a) $2\int_0^2 \pi[4^2-(x^2)^2]dx = 51.2\pi \approx 160.8495$

(b) $\int_0^4 \pi(\sqrt{y})^2dy = 8\pi \approx 25.1327$

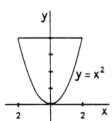

$y = x^2$

12. The points of intersection are $(4,0)$ & $(1,3)$.

$$\text{Volume} = \int_0^3 \pi[(4-y)^2 - (y-2)^4]dy$$

$$= 14.4\pi \approx 45.2389$$

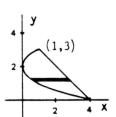

15. Let $15 + \dfrac{19}{x+1} = \dfrac{x}{3} + 10$.

$x^2 - 14x - 72 = 0$; $(x-18)(x+14) = 0$; $x = 18$

Then $p = \frac{1}{3}(18) + 10 = 16$, so the equilibrium price is \$16,000.

18. $CS = \int_0^{\sqrt{3.5}}[(9-x^2) - 5.5]dx = \frac{7}{3}\sqrt{3.5} \approx 4.365$

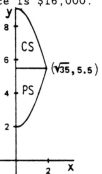

Therefore, CS is \$4,365.

Therefore, PS is \$4,365 (by symmetry).

21. (a) Let $P'(x) = 100 + 5x^{1/2}$ be the rate of increase of the population.

$P(0) = 10000$, so $P(x) = 100x + \frac{10}{3}x^{3/2} + 10000$.

$\therefore P(9) = 900 + 90 + 10000 = 10,990$ persons 9 years from now.

(b) The average yearly increase is $\dfrac{10990 - 10000}{9} = 110$ persons/year.

24. $8400 = 4000(1.0425)^{-6} + x(1.0425)^{-14}$

$\therefore x = 9462.896$, so the second payment will be \$9,462.90.

$$\begin{array}{ccc} & 4,000 & x \\ \hline 0 & 6 & 14 \quad n \end{array}$$

27. (a) $1 = \int_0^8 kx^{1/3}dx = \left[\frac{3k}{4}x^{4/3}\right]_0^8 = 12k$, so $k = \frac{1}{12}$.

(b) $\text{prob}\{0 \le X \le 1\} = \int_0^1 \frac{1}{12}x^{1/3}dx = \frac{1}{16} = 0.0625$

(c) $\mu = \int_0^8 xf(x)dx = \int_0^8 \frac{1}{12}x^{4/3}dx = \frac{32}{7}$

30. $\text{prob}\{0 \leq T \leq 2\} = \int_0^2 0.03 e^{-0.03t} dt = -e^{-0.06} + 1 \approx 0.0582$

33. $1 = \int_a^\infty 27x^2(1+x^3)^{-2} dx = \lim_{b \to \infty} \left[-9(1+x^3)^{-1} \right]_a^b = 9(1+a^3)^{-1}$, so $a = 2$.

36. $\text{Work} = \int_1^8 5x^{2/3} dx = 93$ foot-pounds.

Problem Set 10.1   Functions of Two or More Variables

3. (a) $\dfrac{1}{256}$ (b) $\dfrac{2}{16} - \dfrac{1}{8}$ (c) $\dfrac{x}{16x^4} - \dfrac{1}{16x^3}$ (d) $\dfrac{x+h}{y^4} - \dfrac{x}{y^4} = \dfrac{h}{y^4}$

6. (a) $0+16-5 = 11$ (b) $0+4-5 = -1$ (c) $x^2 \ln(2x^2-3) + 4x^2 - 5$

   (d) $\left[(x+h)^2 \ln[(x+h)y-3] + y^2 - 5\right] - \left[x^2 \ln(xy-3) + y^2 - 5\right]$

   $= (x+h)^2 \ln(xy+hy-3) - x^2 \ln(xy-3)$

9. (a) $\dfrac{0-2}{0-2} = 1$ (b) $\dfrac{6-12}{12-6} = -1$

   (c) $\dfrac{-2z^2-2z}{z^2+4z} = \dfrac{-2(z+1)}{z+4}$ $(z \ne 0)$ (d) $\dfrac{yz-2x}{xy-2z}$

12. (a) e^5 (b) $10e^{23}$ (c) $(z^2+z)e^0 = (z^2+z)$ (d) $(x+y^2)e^{4x-2y-z}$

15. (a) $V(x,y) = x^2y$; $S(x,y) = 2x^2+4xy$
    (b) $V(2x,\tfrac{1}{2}y) = (2x)^2(\tfrac{1}{2}y) = 2x^2y = 2[x^2y]$, so V is doubled.
    (c) $C(x,y) = 2x^2(0.30) + 4xy(0.18) = 0.60x^2 + 0.72xy$

18. (a) $285(200)^{0.78}(200)^{0.22} = 285(200)^1 = 57,000$
    (b) $285(100)^{0.78}(300)^{0.22} \approx 36,292$
    (c) $285(300)^{0.78}(100)^{0.22} \approx 67,143$

21. $x+y+z = 5$ 24. $-x+6y+z = 6$ 27. $z = x^2+y^2-9$

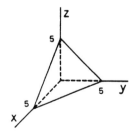

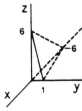
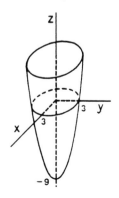

30. $z = x^2+y^2+1$

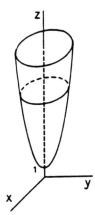

33. $z = 9-x^2-y^2$

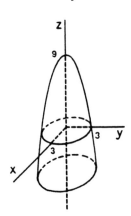

36. $z = x^2+ \frac{1}{9}y^2$

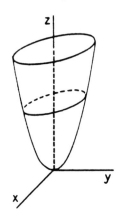

39. $86.4 = k(1800)(40)^2$, so $k = 0.00003$; then $S = 0.00003wr^2$.
Thus, if $w = 2400$ & $r = 60$, then $S = 0.00003(2400)(60)^2 = 259.2$ ft.

42. (a) $3x+y = 9$

(b) $x+4z = 8$

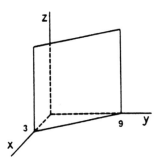

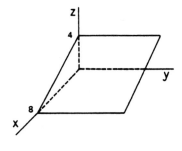

(c) $z = x^2$

(d) $y = (x-1)^2$

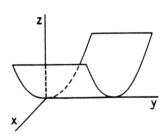

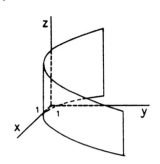

Problem Set 10.2 Partial Derivatives

3. $z_x = y^2 - 8x$; $z_y = 2xy + 3$

6. $z_x = 2xy^{-1/2} - 2x^{-1/3}$; $z_y = -\frac{1}{2}x^2 y^{-3/2}$

9. $z_x = (x^2 y)(2e^{2x}) + (e^{2x})(2xy) = 2xye^{2x}(x+1)$; $z_y = x^2 e^{2x}$

12. $z_x = y^2 e^{x^2 y^{-1}}(2xy^{-1}) = 2xye^{x^2 y^{-1}}$

$z_y = (y^2)\left[\left(e^{x^2 y^{-1}}\right)\left(-x^2 y^{-2}\right)\right] + \left(e^{x^2 y^{-1}}\right)(2y) = e^{x^2 y^{-1}}(2y - x^2)$

15. $z_x = \dfrac{(2x+y)^3}{x^3 y^2} \dfrac{(2x+y)^3(3x^2 y^2) - (x^3 y^2)[3(2x+y)^2(2)]}{(2x+y)^6} = \dfrac{3y}{x(2x+y)}$

$z_y = \dfrac{(2x+y)^3}{x^3 y^2} \dfrac{(2x+y)^3(2x^3 y) - (x^3 y^2)[3(2x+y)^2]}{(2x+y)^6} = \dfrac{4x-y}{y(2x+y)}$

18. $w_z = 4(xyz^3 - z)^3(3xyz^2 - 1)$

21. $f_x(x,y) = 2x + 2y^2$, so $f_x(2,1) = 4 + 2 = 6$.

$f_y(x,y) = 4xy - 1$, so $f_y(2,-1) = -8 - 1 = -9$.

24. $f_x(x,y) = (x^2)[e^{(x-2)y}y] + [e^{(x-2)y}](2x)$, so $f_x(2,1) = 4e^0 + 4e^0 = 8$.

$f_y(x,y) = x^2 e^{(x-2)y}(x-2)$, so $f_y(2,-1) = 0$.

27. $f_x(x,y) = -2x$, so $f_x(2,1) = -4$.

30. $f_x(x,y) = 2xe^{x^2 + 2y^2}$, so $f_x(1,-1) = 2e^3 \approx 40.171$

33. $f(x,y) = 4[\ln(2 + 2x^2 + 2y^2)]^{-1}$
$f_x(x,y) = -4[\ln(2 + 2x^2 + 2y^2)]^{-2}(2 + 2x^2 + 2y^2)]^{-1}(4x)$,
    so $f_x(1,0) = -4(\ln4)^{-2} \approx -2.08$. The grade is about 208%.

36. $R(L,r) = kLr^{-4}$; $R_L(L,r) = kr^{-4}$; $R_r(L,r) = -4kLr^{-5}$

39. (a) $f_x(x,y) = 2y^3 - 4xy$, so $f_{xx}(x,y) = -4y$.

(b) $f_{xy}(x,y) = 6y^2 - 4x$

(c) $f_y(x,y) = 6xy^2 - 2x^2$, so $f_{yx}(x,y) = 6y^2 - 4x$

42. $f_x(x,y) = 4xy(x^2+y)^{-1}$ & $f_y(x,y) = 2y(x^2+y)^{-1} + 2\ln(x^2+y)$

   (a) $f_{xx}(x,y) = 4xy[-(x^2+y)^{-2}(2x)] + (4y)[(x^2+y)^{-1}] = 4y(y-x^2)(x^2+y)^{-2}$

   (b) $f_{xy}(x,y) = (4xy)[-(x^2+y)^{-2}] + (4x)[(x^2+y)^{-1}] = 4x^3(x^2+y)^{-2}$

   (c) $f_{yx}(x,y) = -2y(x^2+y)^{-2}(2x) + 2(x^2+y)^{-1}(2x) = 4x^3(x^2+y)^{-2}$

45. $\dfrac{\partial X_A}{\partial p_B} = -250;\ \dfrac{\partial X_B}{\partial p_A} = -420.$ Therefore, A & B are complementary.

48. $\dfrac{\partial X_A}{\partial p_B} = 48;\ \dfrac{\partial X_B}{\partial p_A} = -60.$ Neither complementary nor competitive.

51. $T_x(x,y) = -x$, so $T_x(4,2) = -4$. Starting at the point 4 feet from the left edge and 2 feet from the bottom edge, and moving Δx feet while staying 2 feet from the bottom edge results in about a $(4\Delta x)°$C drop in temperature.

   $T_y(x,y) = -3y$, so $T_y(4,2) = -6$. Starting at the point 4 feet from the left edge and 2 feet from the bottom edge, and moving Δy feet while staying 4 feet from the left edge results in about a $(6\Delta y)°$C drop in temperature.

54. $A_r(P,r,t) = Pte^{rt}$, so $A_r(1000,0.07,10) = 10000e^{0.7} = \$20,137.53$.
   If the rate changes by Δr while keeping the principal at $1000 and the time at 10 years, the amount will change by about $\$20,137.53\Delta r$. For example, if r changes to 8% so that Δr is 1% then the accumulated amount will increase by about 1% of $20,137.53 which is about $201.37.

57. $V = \pi r^2 h$, $dV = V_r dr + V_h dh = 2\pi rhdr + \pi r^2 dh = \pi r(2hdr + rdh)$

   For $r = 10$, $dr = 0.3$, $h = 30$, & $dh = 0.4$, $dV = 10\pi(1.8 + 4) = 220\pi$, so the volume increases by about $220\pi \approx 691$ cubic centimeters.

3. $f_x(x,y) = 2(x+1)+4 = 2x+6$; $f_{xx}(x,y) = 2$; $f_{xy}(x,y) = 0$;

$f_y(x,y) = (2(y-4)-2 = 2y-10$; $f_{yy}(x,y) = 2$, so $T(x,y) = (2)(2)-(0)^2 = 4$.

$2x+6 = 0$ & $2y-10 = 0$ if $x = -3$ & $y = 5$, so $(-3,5)$ is the only critical point. Since $T(-3,5) = 4 > 0$ & $f_{xx}(-3,5) = 2 > 0$, $(-3,5)$ gives a local minimum.

6. $f_x(x,y) = 6x-y+5$; $f_{xx}(x,y) = 6$; $f_{xy}(x,y) = -1$;

$f_y(x,y) = -x+2y+1$; $f_{yy}(x,y) = 2$, so $T(x,y) = (6)(2)-(-1)^2 = 11$.

$6x-y+5 = 0$ & $-x+2y+1 = 0$ if $x = -1$ & $y = -1$, so $(-1,-1)$ is the only critical point. Since $T(-1,-1) = 11 > 0$ & $f_{xx}(-1,-1) = 6 > 0$, $(-1,-1)$ gives a local minimum.

9. $f_x(x,y) = 3x^2+4xy+4$; $f_{xx}(x,y) = 6x+4y$; $f_{xy}(x,y) = 4x$;

$f_y(x,y) = 2x^2-8$; $f_{yy}(x,y) = 0$, so $T(x,y) = (6x+4y)(0)-(4x)^2 = -16x^2$.

$3x^2+4xy+4 = 0$ & $2x^2-8 = 0$ if $x = 2$ & $y = -2$, or $x = -2$ & $y = 2$, so $(2,-2)$ and $(-2,2)$ are the only critical points.
Since $T(2,-2) = -64 < 0$ & $T(-2,2) = -64 < 0$, $(2,-2)$ & $(-2,2)$ give saddle points.

12. $f_x(x,y) = 3x^2-3$; $f_{xx}(x,y) = 6x$; $f_{xy}(x,y) = 0$;

$f_y(x,y) = 3y^2-12$; $f_{yy}(x,y) = 6y$, so $T(x,y) = (6x)(6y)-(0)^2 = 36xy$.

$3x^2-3 = 0$ & $3y^2-12 = 0$ if $x = \pm1$ & $y = \pm2$, so $(1,2)$, $(-1,2)$, $(1,-2)$, & $(-1,-2)$ are the only critical points.
$T(1,-2)$ & $T(-1,2)$ are negative so $(1,-2)$ & $(-1,2)$ give saddle points.
$T(1,2) = 72$ & $f_{xx}(1,2) = 6 > 0$, so $(1,2)$ gives a local minimum.
$T(-1,-2) = 72$ & $f_{xx}(-1,-2) = -6 < 0$, so $(-1,-2)$ gives a local maximum.

15. $256 = xyz$ & $S = xy + 2xz + 2yz$
$S(x,y) = xy + 2(256y^{-1}) + 2(256x^{-1})$

$S_x(x,y) = y-512x^{-2}$; $S_{xx}(x,y) = 1024x^{-3}$

$S_{xy}(x,y) = 1$

$S_y(x,y) = x-512y^{-2}$; $S_{yy}(x,y) = 1024y^{-3}$

$T(x,y) = (1024x^{-3})(1024y^{-3}) - (1)^2$

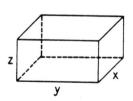

15. (cont.) $y-512x^{-2} = 0$ & $x-512y^{-2} = 0$ if $x = 8$ & $y = 8$, so $(8,8)$ is the only critical point.

$T(8,8) = 3 > 0$ & $S_{xx}(8,8) = 2 > 0$, so $(8,8)$ gives a local minimum which is the global minimum. The base of the tank should be 8' by 8' & the depth should be 4' in order to use the least amount of material.

18. $xyz = 8$ & $S = 4xy + 2yz + xz$
$S(x,y) = 4xy + 2(8x^{-1}) + (8y^{-1})$

$S_x(x,y) = 4y-16x^{-2}$; $S_{xx}(x,y) = 32x^{-3}$

$S_{xy}(x,y) = 4$

$S_y(x,y) = 4x-8y^{-2}$; $S_{yy}(x,y) = 16y^{-3}$

$T(x,y) = (16x^{-3})(16y^{-3}) - (4)^2$

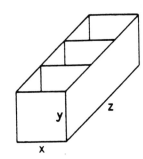

$4y-16x^{-2} = 0$ & $4x-8y^{-2} = 0$ if $x = 2$ & $y = 1$, so $(2,1)$ is the only critical point.

$T(2,1) = 16 > 0$ & $S_{xx}(2,1) = 4 > 0$, so $(2,1)$ gives a local minimum which is the global minimum. To use the least amount of material, the base should be 2' by 4' and the height should be 1', with the two partitions placed perpendicular to the longer sides.

21. $P(x,y) = (80-2.5x)x + (120-4y)y - (30x+40y-4xy)$
$= -2.5x^2-4y^2+4xy+50x+80y$

$P_x(x,y) = -5x+4y+50$; $P_{xx}(x,y) = -5$; $P_{xy}(x,y) = 4$

$P_y(x,y) = -8y+4x+80$; $P_{yy}(x,y) = -8$

$-5x+4y+50 = 0$ & $-8y+4x+80 = 0$ if $x = 30$ & $y = 25$, so $(30,25)$ is the only critical point.

$T(x,y) = (-5)(-8) - (4)^2 = 24 > 0$ & $P_{xx}(x,y) = -5 < 0$, so $(30,25)$ gives a local maximum, which is the global maximum. Thus, profit is maximized if 30 tricycles of the first kind and 25 of the second kind are made each day.

24. $xyz = 24$. Minimize $W = 2x+2y+z$ if $x \geq 0$, $y \geq 0$, $z \geq 0$.

$W(x,y) = 2x+2y+24x^{-1}y^{-1}$

$W_x(x,y) = 2-24x^{-2}y^{-1}$; $W_{xx}(x,y) = 48x^{-3}y^{-1}$; $W_{xy}(x,y) = 24x^{-2}y^{-2}$

$W_y(x,y) = 2-24x^{-1}y^{-2}$; $W_{xx}(x,y) = 48x^{-1}y^{-3}$

24. (cont.) $2-24x^{-2}y^{-1} = 0$ & $2-24x^{-1}y^{-2} = 0$ if $x = \sqrt[3]{12}$, $y = \sqrt[3]{12}$; $(\sqrt[3]{12},\sqrt[3]{12})$
is the only critical point.

$T(x,y) = (48x^{-3}y^{-1})(48x^{-1}y^{-3}) - (24x^{-2}y^{-2})^2 = 1728x^{-4}y^{-4}$

Since $T(x,y) > 0$ & $W_{xy}(x,y) > 0$ at the critical point, the critical
point gives a local minimum. The three positive numbers we seek are
$x = \sqrt[3]{12}$, $y = \sqrt[3]{12}$, $z = 2\sqrt[3]{12}$.

27. $R(x,t) = (40x-4x^2)te^{-2t}$

$R_x(x,t) = (40-8x)te^{-2t}$; $R_{xx}(x,t) = -8te^{-2t}$

$R_{xt}(x,t) = -8[-2te^{-2t} + e^{-2t}] = 8e^{-2t}(2t-1)$

$R_t(x,t) = (40x-4x^2)(-2te^{-2t}+e^{-2t}) = 4x(10-x)e^{-2t}(1-2t)$

$R_{tt}(x,t) = (40x-4x^2)(4te^{-2t}-2e^{-2t}-2e^{-2t}) = -16x(10-x)e^{-2t}(1-t)$

$(40-8x)te^{-2t} = 0$ & $4x(10-x)e^{-2t}(1-2t) = 0$ if
$\qquad\qquad\qquad\qquad$ ($x=5$ or $t=0$) & ($x=0$ or $=10$ or $t=\frac{1}{2}$)

Clearly, $x=0$, $x=10$, & $t=0$ each gives the minimum value of 0 for R, so
$(5,\frac{1}{2})$ is the only critical point we need to consider.

$T(x,y) = (-8te^{-2t})[-16x(10-x)e^{-2t}(1-t)] - [8e^{-2t}(2t-1)]^2$

$T(5,\frac{1}{2}) = 800e^{-2} > 0$ & $R_{xx}(5,\frac{1}{2}) = -40e^{-1} < 0$, so $(5,\frac{1}{2})$ gives a local
maximum which is the global maximum. That maximum reaction is

$R(5,\frac{1}{2}) = 50e^{-1} \approx 18.4$ (appropriate units).

Problem Set 10.4 Constrained Optimization and Lagrange's Method

3. Let $F(x,y,\lambda) = (2x^2-xy+4y^2+100) + \lambda(2x+5y-76)$.

If $F_x = 4x-y+2\lambda = 0$, $F_y = -x+8y+5\lambda = 0$, and $F_\lambda = 2x+5y-76 = 0$,
then $x = 10.5$ & $y = 11$. The minimum value is $f(10.5,11) = 689$.

6. Let $F(x, y, \lambda) = x^2y + \lambda(x^2+y^2-20)$.
   If $F_x = 2xy+2x\lambda = 0$, $F_y = x^2+2y\lambda = 0$, and $F_\lambda = x^2+y^2-20 = 0$,

   then $x = \pm\sqrt{40/3}$ & $y = \pm\sqrt{20/3}$.

   Maximum is $f(\sqrt{40/3}, \sqrt{20/3}) = f(-\sqrt{40/3}, \sqrt{20/3}) = (40/3)\sqrt{20/3} \approx 34.427$

   Minimum is $f(\sqrt{40/3}, -\sqrt{20/3}) = f(-\sqrt{40/3}, -\sqrt{20/3}) = -(40/3)\sqrt{20/3} \approx -34.427$

9. Let $F(x, y, z, \lambda) = (x^2+y^2+z^2) + \lambda(2x+y+3z-14)$.

   If $F_x = 2x+2\lambda = 0$, $F_y = 2y+\lambda = 0$, $F_z = 2z+3\lambda$, and $F_\lambda = 2x+y+3z-14 = 0$,
   then $x = 2$, $y = 1$, $z = 3$. Minimum value is $f(2,1,3) = 14$.

12. Let $F(x, y, z, \lambda) = xyz + \lambda(x^2+y^2+z^2-3)$
    If $F_x = yz+2x\lambda = 0$, $F_y = xz+2y\lambda = 0$, $F_z = xy+2z\lambda = 0$,
    & $F_\lambda = x^2+y^2+z^2-3 = 0$
    then $x = \pm1$, $y = \pm1$, $z = \pm1$. Minimum value is $f(-1,-1,-1) = -1$,
    which also occurs at $(-1,1,1)$, $(1,-1,1)$ and $(1,1,-1)$.

15. Let $F(x, y, \lambda) = 300x^{2/3}y^{1/3} + \lambda(400x+500y-180000)$

    If $F_x = 200x^{-1/3}y^{1/3}+400\lambda = 0$, $F_y = 100x^{2/3}y^{-2/3}+500\lambda = 0$,
    & $F_\lambda = 400x+500y-180000 = 0$
    then $x = 300$, $y = 120$, $\lambda = -0.5(300)^{-1/3}(120)^{1/3} \approx -0.368$.

    For maximum production use 300 units of labor & 120 units of capital.
    The marginal productivity of money is 0.368.

18. The surface area is $S(r,h) = 2\pi rh+2\pi r^2$, subject to $\pi r^2h = 1000$.

    Let $F(r, h, \lambda) = (2\pi rh+2\pi r^2) + \lambda(\pi r^2h-1000)$.

    If $F_r = 2\pi h+4\pi r+2\pi rh\lambda = 0$, $F_h = 2\pi r+\pi r^2\lambda = 0$, and $F_\lambda = \pi r^2h-1000 = 0$,

    then $r = (500/\pi)^{1/3} \approx 5.42$ cm., and $h = 2(500/\pi)^{1/3} \approx 10.84$ cm.

21. Let $F(x, y, z, \lambda) = (12x+14y+20z) + \lambda(x^2+y^2+z^2-740)$, $x \geq 0$, $y \geq 0$, $z \geq 0$.

    If $F_x = 12+2x\lambda = 0$, $F_y = 14+2y\lambda = 0$, $F_z = 20+2z\lambda = 0$,
    & $F_\lambda = x^2+y^2+z^2-740 = 0$,
    then x=12, y=14, z=20 for max. daily profit of $P(12,14,20) = \$740$.

114   Problem Set 10.4

24. Let $F(x,y,\lambda) = kx^ay^b + \lambda(cx+dy-M)$. Let (1) $F_x = akx^{a-1}y^b + c\lambda = 0$,

$\qquad\qquad\qquad\qquad\qquad\qquad\qquad$ (2) $F_y = bkx^ay^{b-1} + d\lambda = 0$,

$\qquad\qquad\qquad\qquad\qquad\qquad\qquad$ (3) $F_\lambda = cx+dy-M = 0$,

Then (4) $\lambda = (-\frac{1}{c})(akx^{a-1}y^b)$ [from (1)]

$\qquad$ (5) $x^{a-1}y^{b-1}(-ckbx + dkay) = 0$ [substituting (4) into (2) and
$\qquad\qquad\qquad\qquad\qquad\qquad\qquad\qquad\qquad\qquad\qquad\qquad$ simplifying]

$\qquad$ (6) $y = (\frac{bc}{ad})x$ [from (5)]

$\qquad$ (7) $acx + bcx = aM$ [substituting (6) into (3) and simplifying]

$\qquad$ (8) $x = (\frac{a}{c})M$ [from (7) with using $a+b = 1$]

$\qquad$ (9) $y = (\frac{b}{d})M$ [substituting (8) into (6) and simplifying]

Problem Set 10.5 Application: The Least Squares Line

3. (i) $S = (-1m+b-1)^2 + (1m+b-2)^2 + (2m+b-2)^2 + (5m+b-3)^2$

$\qquad$ (ii) Let $S_m = 2(-1m+b-1)(-1) + 2(1m+b-2) + 2(2m+b-2)(2) + 2(5m+b-3)(5)$
$\qquad\qquad\quad = 2(31m+7b-10) = 0$
$\qquad\qquad$ Let $S_b = 2(-1m+b-1) + 2(1m+b-2) + 2(2m+b-2) + 2(5m+b-3)$
$\qquad\qquad\qquad = 2(7m+4b-8) = 0$

$\qquad$ (iii) $m = 0.32$, $b = 1.44$, so $y = 0.32x + 1.44$

6. $m = \dfrac{(14)(8) - 4(25)}{(14)^2 - 4(62)} = \dfrac{-3}{13} \approx -0.231$

$\qquad b = \dfrac{(8) - (-3/13)(14)}{4} = \dfrac{73}{26} \approx 2.808$

$\qquad$ Thus, $y = -0.231x + 2.808$ (approximately).

x	y	xy	x^2
1	3	3	1
3	2	6	9
4	1	4	16
6	2	12	36
14	8	25	62

9. (a) $m = \dfrac{(75)(19.3) - 6(305.5)}{(75)^2 - 6(1375)} = \dfrac{-385.5}{-2625} \approx 0.147$

$\qquad b \approx \dfrac{(19.3) - (0.147)(75)}{6} \approx 1.38$

$\qquad$ Thus, $y = 0.147x + 1.38$ (approximately).

x	y	xy	x^2
0	1.8	0	0
5	2.1	10.5	25
10	2.6	26	100
15	3.0	45	225
20	4.2	84	400
25	5.6	140	625
75	19.3	305.5	1375

$\qquad$ (b) If $x = 18$, $y \approx 0.147(18) + 1.38 \approx \$4.0$ thousand per capita.
$\qquad$ (c) If $x = 40$, $y \approx 0.147(40) + 1.38 \approx \$7.3$ thousand per capita.

12. $m = \dfrac{(28)(156) - 7(741)}{(28)^2 - 7(140)} = \dfrac{-819}{-196} \approx 4.18$

$b \approx \dfrac{(156) - (4.18)(28)}{7} \approx 5.57$

Thus, $y = 4.18x + 5.57$ (approximately).

If $x = 9$, $y = 4.18(9) + 5.57 \approx 43$ new cases.

x	y	xy	x^2
1	10	10	1
2	15	30	4
3	18	54	9
4	21	84	16
5	24	120	25
6	33	198	36
7	35	245	49
28	156	741	140

15.

x	y	Y ℓny	xY	x^2
1	3.1	$\ell n(3.1)$	$\ell n(3.1)$	1
2	6.3	$\ell n(6.3)$	$2\ell n(6.3)$	4
3	11.9	$\ell n(11.9)$	$3\ell n(11.9)$	9
4	24.1	$\ell n(24.1)$	$4\ell n(24.1)$	16
5	47.8	$\ell n(47.8)$	$5\ell n(47.8)$	25
15		12.498	44.306	55

$m \approx \dfrac{(15)(12.498) - 5(44.306)}{(15)^2 - 5(55)} \approx 0.6813$

$b \approx \dfrac{(12.498) - 0.6813(15)}{5} \approx 0.4561$, so $a = e^b \approx e^{0.4561} \approx 1.5779$

Thus, $y = 1.58e^{0.681x}$ (approximately).

Chapter 10 Review Problem Set

3. $f(r,h) = 4\pi r^2 + 4\pi(r+h)^2 = 4\pi(2r^2 + 2rh + h^2)$
   $f(3, 0.5) = 4\pi(18 + 3 + 0.25) = 85\pi \approx 267.034$

6. $V = \frac{1}{2}(a+b)dh$

9. (a) $f_x(x,y) = 4x-3y$; $f_y(x,y) = -3x+3y^2$, so $f_y(1,0) = -3$.

(b) $f_x(x,y) = (x^2)\dfrac{2}{2x+3y} + [\ln(2x+3y)](2x) = 2x^2(2x+3y)^{-1} + 2x\ln(2x+3y)$.

$f_y(x,y) = (x^2)\dfrac{3}{2x+3y}$, so $f_y(1,0) = \dfrac{3}{2}$.

(c) $f_x(x,y) = e^{x^2 y}(2xy)$; $f_y(x,y) = e^{x^2 y}(x^2)$, so $f_y(1,0) = e^0 = 1$.

(d) $f_x(x,y) = 4(3x-2x^{-1}y)^3(3+2x^{-2}y)$;
$f_y(x,y) = 4(3x-2x^{-1}y)^3(-2x^{-1}) = -8x^{-1}(3x-2x^{-1}y)^3$, so $f_y(1,0) = -216$.

12. $z_x(x,y) = 2x-3y$, so the desired slope is $z_x(3,1) = 6-3 = 3$.

15. $f_P(P,r) = (1+r)^{20}$, so $f_P(2000,0.08) = (1.08)^{20} = \$4.66$. If r remains at 8% and P changes by ΔP, then the accumulated amount changes by about $\$4.66\Delta P$. For example, if $2,001 is deposited, so that $\Delta P = 1$, then the accumulated amount will increase by about $4.66.

$f_r(P,r) = 20P(1+r)^{19}$, so $f_r(2000,0.08) = 40000(1.08)^{19} = \$172,628.04$.

If P remains at $2,000 and r changes by Δr, then the accumulated amount changes by about $\$172628.04\Delta r$. For example, if r increases to 9%, so that $\Delta r = 0.01$, then the accumulated amount will increase by about $1,726.28.

18. $f_x(x,y) = 3x^2-9y$ and $f_y(x,y) = 3y^2-9x$.

$3x^2-9y = 0$ and $3y^2-9x = 0$ at $(0,0)$ and $(3,3)$.

$T(x,y) = f_{xx}(x,y)f_{yy}(x,y) - [f_{xy}(x,y)]^2$
$= (6x)(6y) - (-9)^2 = 36xy-81$

$T(0,0) = -81 < 0$, so $(0,0)$ gives a saddle point.
$T(3,3) = (18)(18)-81 > 0$ and $f_{xx}(3,3) = 18 > 0$, so $(3,3)$ gives a local minimum value.

21. Let R be the weekly revenue.

Then $R(p,q) = (300 - 0.5p + 0.3q)p + (400 + 0.2p - 0.4q)q$

$$= 300p - 0.5p^2 + 0.5pq + 400q + 0.2pq - 0.4q^2$$

$R_p(p,q) = 300 - p + 0.5q$ and $R_q(p,q) = 0.5p + 400 - 0.8q$.

$300 - p + 0.5q = 0$ and $0.5p + 400 - 0.8q = 0$ at $(800,1000)$.

$T(p,q) = T_{pp}(p,q)T_{qq}(p,q) - [T_{pq}(p,q)]^2$

$$= (-1)(-0.8) - (0.5)^2 > 0 \text{ and } R_{pp}(800,1000) = -0.8 < 0, \text{ so}$$

$(800,1000)$ gives a local minimum which is the global minimum.

Sell the mowers for p = $800 and q = $1,000.

24. Let $F(x,y,\lambda) = x^2y + \lambda(x^2+y^2-27)$.

Let $F_x(x,y,\lambda) = 2xy+2x\lambda = 0$,

$F_y(x,y,\lambda) = x^2+2y\lambda = 0$, and

$F_\lambda(x,y,\lambda) = x^2+y^2-27 = 0$. Then $x = \pm3\sqrt{2}$ and $y = \pm3$.

$f(3\sqrt{2},3) = f(-3\sqrt{2},3) = 54$ is the maximum value of $f(x,y)$.

27. (a) $1400 = k[(2)(1)^2/20]$, so $k = 14000$

(b) $L(w,d,s) = 14000wd^2s^{-1}$, so $L_s(w,d,s) = -14000wd^2s^{-2}$.

Then $L_s(2,1,20) = -70$, so the load it will safely support will decrease by about 70 pounds, to 1330 pounds.

Problem Set 11.1   Double Integrals over Rectangles

3. $\left[y^2 \ell nx\right]_{x=2}^{5} = y^2(\ell n5 - \ell n2) = y^2 \ell n(2.5)$

6. $\int_0^2 14x\,dx = \left[7x^2\right]_0^2 = 28$

9. $\int_0^3 \int_0^1 (x^2+3y^2)\,dx\,dy = \int_0^3 \left[\frac{1}{3}x^3 + 3y^2 x\right]_{x=0}^{1} dy = \int_0^3 (\frac{1}{3}+3y^2)\,dy = \left[\frac{1}{3}y + y^3\right]_0^3 = 28$

12. $\int_0^1 \int_1^7 ye^{2x}\,dy\,dx = \int_0^1 \left[\frac{1}{2}y^2 e^{2x}\right]_{y=1}^{7} dx = \int_0^1 24e^{2x}\,dx = \left[12e^{2x}\right]_0^1 = 12(e^2-1)$

15. $\int_0^2 \int_0^{\ell n3} xe^y\,dy\,dx = \int_0^2 \left[xe^y\right]_{y=0}^{\ell n3} dx = \int_0^2 (3x-x)\,dx = \left[x^2\right]_0^2 = 4$

18. $\int_1^2 \int_0^1 xye^{xy^2}\,dy\,dx = \int_1^2 \left[\frac{1}{2}e^{xy^2}\right]_{y=0}^{1} dx = \int_1^2 \frac{1}{2}(e^x-1)\,dx = \left[\frac{1}{2}(e^x-x)\right]_1^2 = \frac{1}{2}(e^2-e-1)$

21. $\int_0^1 \int_0^1 4x^3(x^2y+1)^3\,dy\,dx = \int_0^1 \left[x(x^2y+1)^4\right]_0^1 dx = \int_0^1 [x(x^2+1)^4 - x]\,dx$

$= \left[\frac{1}{10}(x^2+1)^5 - \frac{1}{2}x^2\right]_0^1 = 2.6$

24. $\left[\frac{1}{4}e^{x^2y}\right]_{x=0}^{3} = \frac{1}{4}(e^{18y}-1)$

27. $\int_2^3 \left[\frac{-(2x+1)}{(x+2)(x-1)y}\right]_{y=2}^{4} dx = \frac{1}{4}\int_2^3 \frac{2x+1}{(x+2)(x-1)}\,dx = \frac{1}{4}\int_2^3 \left[\frac{1}{x+2} + \frac{1}{x-1}\right]dx$

$= \frac{1}{4}\left[\ell n(x+2) + \ell n(x-1)\right]_2^3 = \frac{1}{4}\ell n(2.5)$

30. $\int_0^1 \left[\frac{1}{2}e^{x^2+y}\right]_{x=0}^{2} dy = \frac{1}{2}\int_0^1 [e^{4+y} - e^y]\,dy = \frac{1}{2}\left[e^{4+y} - e^y\right]_0^1 = \frac{1}{2}(e^5 - e^4 - e + 1)$

33. $\left[\frac{2}{3}x^{3/2}\right]_4^9 \left[\frac{1}{3}y^3\right]_0^1 \left[\frac{1}{4}z^4\right]_0^2 = \left(\frac{38}{3}\right)\left(\frac{1}{3}\right)(4) = \frac{152}{9}$

Problem Set 11.2 Double Integrals over General Regions

3. $\displaystyle\int_0^2 \left[x^3+x^2y\right]_{x=1}^{2y} dy = \int_0^2 (12y^3-y-1)dy = \left[3y^4-\frac{1}{2}y^2-y\right]_0^2 = 44$

6. $\displaystyle\int_0^1 \left[e^{x^2}\ell ny\right]_{y=1}^{e^x} dx = \int_0^1 xe^{x^2}dx = \left[\frac{1}{2}e^{x^2}\right]_0^1 = \frac{1}{2}(e-1)$

9. $\displaystyle\int_0^2\int_{y^2-y}^y 2x\ dxdy = \int_0^2 (-y^4+2y^3)dy = 1.6$

12. $\displaystyle\int_0^2\int_0^{(4-x^2)^{1/2}} 2xy\ dydx = \int_0^2 (4x-x^3)dx = 4$

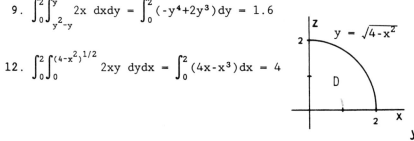

15. $\displaystyle\int_{-1}^1\int_{x^2}^{x+2} 4y\ dydx = \int_{-1}^1 (2x^2+8x+8-2x^4)dx = \frac{248}{15}$

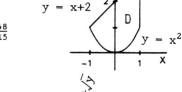

18. $\displaystyle\int_1^4\int_2^{2+y^{1/2}} 2xy^{-1/2}dxdy = \int_1^4 (4+y^{1/2})dy = \frac{50}{3}$

21. $\displaystyle\int_1^2\int_{2x}^4 2x^{-1}y\ dydx = \int_1^2 (16x^{-1}-4x)dx = 16\ell n2 - 6$

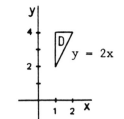

24. $\int_0^1 \int_y^{y^{1/3}} f(x,y)\,dxdy$

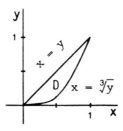

27. $\int_{-1}^0 \int_{2x^3}^0 x\,dydx \;+\; \int_0^2 \int_0^{2x^3} x\,dydx$

$= \int_{-1}^0 -2x^4\,dx \;+\; \int_0^2 2x^4\,dx$

$= -0.4 + 12.8 = 12.4$

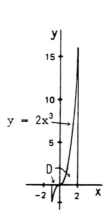

30. The point of intersection is $(1,e)$.

Area $= \int_0^1 \int_{e^x}^{2e-e^x} 1\,dydx = \int_0^1 (2e-2e^x)\,dx = 2$

33. $\int_0^1 \int_0^{e^x} (x+2)\,dydx = \int_0^1 (x+2)e^x\,dx = 2e-1$

[In the last integral use integration by parts with $u = x+2$ and $dv = e^x\,dx$.]

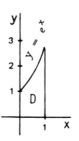

36. Points of intersection are $(-2,4)$ and $(1,1)$.

$\int_{1/2}^1 \int_{-y^{1/2}}^{y^{1/2}} x^2y\,dxdy \;+\; \int_1^4 \int_{-y^{1/2}}^{2-y} x^2y\,dxdy$

$= \int_{1/2}^1 \tfrac{2}{3}y^{5/2}\,dy \;+\; \int_1^4 \tfrac{1}{3}(8y-12y^2+6y^3-y^4+y^{5/2})\,dy$

$= \dfrac{3186 - 5\sqrt{2}}{420} \;+\; \dfrac{1553}{210} \approx 7.5689$

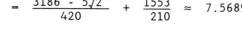

3. $\int_0^3 \int_{2x-6}^0 \frac{1}{2}(6-2x+y)\,dy\,dx$

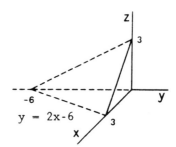
$y = 2x-6$

$= \frac{1}{2}\int_0^3 \frac{1}{2}(2x-6)^2\,dx = 9$

By Formula: $\frac{1}{3}[\frac{1}{2}(6)(3)](3) = 9$

6. $\int_0^2 \int_0^x (4-x^2)\,dy\,dx = \int_0^2 (4x-x^3)\,dx = 4$

9. $\int_0^6 \int_0^{2x/3} (x+2y)\,dy\,dx = \int_0^6 \frac{10}{9}x^2\,dx = 80$

12. $\int_0^4 \int_0^{x^{1/2}} 10x^{1/2}y\,dy\,dx = \int_0^4 5x^{3/2}\,dx = 64$

15. $16000\int_{-2}^2 \int_{-2}^2 e^{-|x|/2}e^{-|y|}\,dx\,dy = 64000\int_0^2 e^{-x/2}\int_0^2 e^{-y}\,dx\,dy$

$= 64000(-2e^{-1}+2)(-e^{-2}+1) \approx 69961$ people.

18. $\int_1^4 \int_0^4 x^{1/2}y^{3/2}\,dx\,dy = \left[\frac{2}{5}y^{5/2}\right]_1^4 \left[\frac{2}{3}x^{3/2}\right]_0^4 = \left(\frac{62}{5}\right)\left(\frac{16}{3}\right) = \frac{992}{15}$

Area of D is 12, so average value of $f(x,y)$ is $\frac{992/15}{12} = \frac{248}{45} \approx 5.5111$.

21. $\int_0^2 \int_0^3 (x^2+y^2)\,dx\,dy = \int_0^2 (9+3y^2)\,dy = 26$, average elevation is $\frac{26}{6} \approx 4.33$ km.

24. $\int_0^3 \int_0^3 (45-x^2-y^2)\,dx\,dy = \int_0^3 (126-3y^2)\,dy = 351$.

Area of the plate is 9, so the average temperature is $351/9 = 39°C$.

27. $\int_0^2 \int_0^8 ye^x\,dx\,dy + \int_0^6 \int_8^{10} ye^x\,dx\,dy = \left[\frac{1}{2}y^2\right]_0^2 \left[e^x\right]_0^8 + \left[\frac{1}{2}y^2\right]_0^6 \left[e^x\right]_8^{10}$

$= 18e^{10}-16e^8-2 \approx 348,779$

30. $M(L) = \int_0^2 \int_{3x}^6 (2x+4y)\,dy\,dx = \int_0^2 (72+12x-24x^2)\,dx = 104$

$\int_0^2 \int_{3x}^6 x(2x+4y)\,dy\,dx = \int_0^2 (72x+12x^2-24x^3)\,dx = 80$, so $\bar{x} = 80/104 \approx 0.77$

$\int_0^2 \int_{3x}^6 y(2x+4y)\,dy\,dx = \int_0^2 (288+36x-45x^3)\,dx = 468$, so $\bar{y} = 468/104 = 4.5$

Chapter 11 Review Problem Set

3. $\int_0^{\ell n3} \int_0^{\ell n2} e^{2x}e^y\,dy\,dx = \left[\tfrac{1}{2}e^{2x}\right]_0^{\ell n3} \left[e^y\right]_0^{\ell n2} = 4$

6. $\int_{1/4}^1 \int_0^{x^{-2}} e^{1/x}\,dy\,dx = \int_{1/4}^1 e^{1/x}x^{-2}\,dx = \left[-e^{1/x}\right]_{1/4}^1 = -e+e^4$

9. $\int_0^3 \int_0^{6-2x} (12-4x-2y)\,dy\,dx = \int_0^3 4(x-3)^2\,dx = 36$

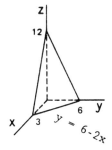

12. $\int_0^4 \int_0^{x^{1/2}} 4xy^3\,dy\,dx = \int_0^4 x^3\,dx = 64$

$\int_0^2 \int_{y^2}^4 4xy^3\,dx\,dy = \int_0^2 (32y^3-2y^7)\,dy = 64$

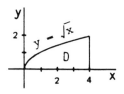

15. $\int_0^1 \int_0^x 2e^{x^3} y \ dydx = \int_0^1 x^2 e^{x^3} dx = \left[\frac{1}{3}e^{x^3}\right]_0^1 = \frac{1}{3}(e-1)$

18. $\int_0^1 \int_0^1 (4-x^2-y^2) dydx = \int_0^1 (\frac{11}{3}-x^2) dx = \frac{10}{3}.$

Area of the square is 1, so the average elevation is $\dfrac{10/3}{1} = \dfrac{10}{3}.$

21. $\int_0^2 \int_{2y}^4 30e^{x-y-2} dxdy = \int_0^2 (30e^{2-y}-30e^{y-2}) dy$

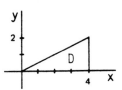

$= 30(e^2+e^{-2}-2)$

The area of the plate is 4, so the average value of $T(x,y)$ is $7.5(e^2+e^{-2}-2) \approx 41.4°C.$

24. $\int_1^e \int_0^{1/x} 4xy \ dydx = \int_1^e 2x^{-1} dx = 2$

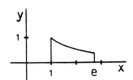

27. $\int_0^1 \int_{x^{1/2}}^{e^x} 2y \ dydx = \int_0^1 (e^{2x}-x) dx = \frac{1}{2}e^2-1$

CHAPTER 12   THE TRIGONOMETRIC FUNCTIONS

Problem Set 12.1   Angles, Arcs, and Triangles

3. 3π (Each half revolution is a π.)

6. $-13\pi/4$ ($-2\pi/4$ for each of the six quadrants passed through plus
   another $-\pi/4$.)

9. $-23\pi/6$ (690° is twenty-three 30°'s.)

12. $1° = \pi/180$ radians, so $-39.4° = -39.4(\pi/180) \approx -0.6877$.

15. $\pi/6 = 30°$, so $7\pi/6 = 7(30°) = 210°$.

18. $\pi/4 = 45°$, so $15\pi/4 = 15(45°) = 675°$.

21. 1 radian $= 180°/\pi$, so 41.23 radians $= 41.23(180°/\pi) \approx 2362.30°$.

24. $s = (2.4)(4) = 9.6$ cm.; $A = \frac{1}{2}(2.4)(4^2) = 19.2$ cm^2.

27. (a) $12.6 = 3.25r$, so $r \approx 3.88$ cm.

   (b) $348° = 348\pi/180$ radians; then $45 = (348\pi/180)r$, so $r \approx 7.41$ in.

30. The minute hand turns 2.5 revolutions, which is $2.5(2\pi) = 5\pi$ radians.
   Then the hour hand turns $5\pi/12$ radians.

33. (a) $\sin(58°) = 300/c$, so $c \approx 353.75$ ft.

   (b) $\tan(58°) = 300/a$, so $a \approx 187.46$ ft.

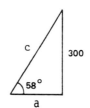

36. $35 = t(14)$, so $t = \frac{35}{14}$ radians. Then $A = \frac{1}{2}(\frac{35}{14})(14^2) = 245$ cm^2.

39. (a) 40 revolutions $= 80\pi$ radians. In one minute the grease spot moves $s = (80\pi)(9) = 720\pi$ in., so the grease spot moves at 720π in/min, which is about 2890.27 in/min.

   (b) In one minute, the belt travels the same distance along each pulley (assuming no slipping).

   That is, $s_{small} = s_{large}$.

   Therefore, $(t_{small})(r_{small}) = (t_{large})(r_{large})$.

$$(t_{small})(5) = (80\pi)(9)$$

$$t_{small} = 144\pi \text{ radians} = 72 \text{ revolutions.}$$

   Thus, the smaller pulley is turning at 72 revolutions per minute.

42. Let D be the other vertex in the figure.

   $\overline{AB} = 20cos(36°)$, so $\overline{BC} = 10cos(36°)$ [since $\overline{AB} = 2\overline{BC}$];

   and $\overline{BD} = 20sin(36°)$.

   Therefore, $x^2 = [20sin(36°)]^2 + [10cos(36°)]^2$; $x \approx 14.27$.

Problem Set 12.2   The Sine and Cosine Functions

3. $r = 2$; $sint = -2/2 = -1$; $cost = 0/2 = 0$.

6. $r^2 = 2+3$, so $r = \sqrt{5}$; $sint = \sqrt{3}/\sqrt{5} = \sqrt{0.6}$; $cost = -\sqrt{2}/\sqrt{5} = -\sqrt{0.4}$.

9. (a) $sin(5\pi/4) = -\sqrt{2}/2$; $cos(5\pi/4) = -\sqrt{2}/2$.

   (b) $sin(-7\pi/4) = \sqrt{2}/2$; $cos(-7\pi/4) = \sqrt{2}/2$.

   (c) $sin(5\pi/2) = 1$; $cos(5\pi/2) = 0$.

   (d) $sin(15\pi/4) = -\sqrt{2}/2$; $cos(15\pi/4) = \sqrt{2}/2$.

12. (a) $\sin(2\pi/3) = \sqrt{3}/2$; $\cos(2\pi/3) = -1/2$.

(b) $\sin(5\pi/3) = -\sqrt{3}/2$; $\cos(5\pi/3) = 1/2$.

(c) $\sin(-7\pi/3) = -\sqrt{3}/2$; $\cos(-7\pi/3) = 1/2$.

(d) $\sin(19\pi/3) = \sqrt{3}/2$; $\cos(19\pi/3) = 1/2$

15. $\sin t > 0$ in quadrants I & II and $\cos t < 0$ in quadrants II & III, so the terminal side of t is in quadrant II.

18. $\sin^2 t = 1 - \cos^2 t = 1 - (-5/13)^2 = 144/169$, so $\sin t = -12/13$.

21. (a) Amplitude is 4.
    Let $0 \le 4t \le 2\pi$. Then $0 \le t \le \pi/2$, so the period is $\pi/2$.

(b) Amplitude is 2.
    Let $0 \le \frac{1}{3}t \le 2\pi$. Then $0 \le t \le 6\pi$, so the period is 6π.

(c) Amplitude is 3.
    Let $0 \le 1.5t \le 2\pi$. Then $0 \le t \le 4\pi/3$, so the period is $4\pi/3$.

24.

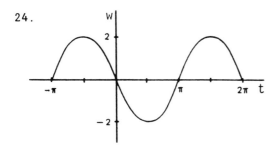

27.

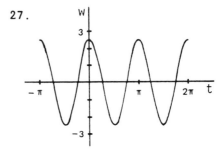

30.

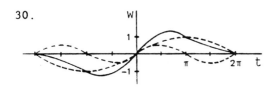

33. Approximately -7.81828

Problem Set 12.2

127

36. (a) 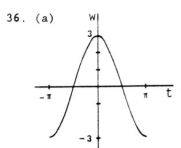 (b) 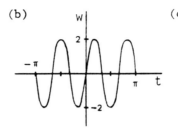 (c)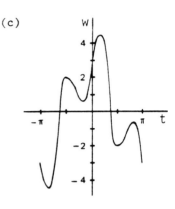

39. (a) g(x) = [x] - x = -(x - [x]). Think of (x - [x]) as the decimal
    part of x, so -(x - [x]) is the negative of the decimal part of x.
    The value of g(x) returns to 0 at each integer value of x, and
    follows the same pattern between integer values. Analytically,
    g(x+1) = [x+1] - (x+1) = [x] + 1 - x - 1 = g(x), and 1 is the
    smallest value of p such that g(x+p) = g(x), for all x.

   (b) h(x) = [2x] - 2x in a similar way. The value of h(x) returns to 0
    at each multiple of 0.5. The period is 0.5. Analytically,
    h(x+0.5) = [2(x+0.5)] - 2(x+0.5) = [2x+1] - 2x - 1
             = [2x] + 1 - 2x - 1 = [2x] - 2x = h(x), and 0.5 is the
    smallest value of p such that h(x+p) = h(x), for all x.

Problem Set 12.3 Derivatives of the Trigonometric Functions

3. $2[-sin(1-3t)](-3) = 6sin(1-3t)$

6. $[-sin(t^3+2t)](3t^2+2) = -(3t^2+2)sin(t^3+2t)$

9. $[2sin(2t^3-1)][cos(2t^3-1)](6t^2) = 12t^2sin(2t^3-1)cos(2t^3-1)$

   $= 6t^2sin[2(2t^3-1)]$

12. $[-sin(sint)](cost) = -cost\ sin(sint)$

15. $\dfrac{(cos4t)(2cos2t) - (sin2t)(-4sin4t)}{cos^24t} = \dfrac{2cos2t\ cos4t + 4sin2t\ sin4t}{cos^24t}$

18. $e^{t+cost}(1-sint)$

21. $f'(x) = (2cos3x)(-sin3x)(3) = -6sin3x\ cos3x = -3sin6x$

    $f''(x) = (-3cos6x)(6) = -18cos6x$

24. $f'(x) = [-sin(x^3)](3x^2) = -3x^2sin(x^3)$

    $f''(x) = (-3x^2)[3x^2cos(x^3)] + [sin(x^3)](-6x) = -9x^4cos(x^3) - 6xsin(x^3)$

27. $\frac{dy}{dx} = -4sin2x - 2sinx\ cosx = -4sin2x - sin2x = -5sin2x$, so the slope is

    $-5(\sqrt{3}/2)$ at the point.

    An equation of the tangent is $y - \frac{1}{4} = -5(\sqrt{3}/2)(x-\frac{\pi}{6})$

30. $\frac{d}{dx}(secx) = \frac{d}{dx}[(cosx)^{-1}] = -(cosx)^{-2}(-sinx) = \frac{1}{cosx}\ \frac{sinx}{cosx} = secx\ tanx$

    $\frac{d}{dx}(cscx) = \frac{d}{dx}[(sinx)^{-1}] = -(sinx)^{-2}(cosx) = \frac{-1}{sinx}\ \frac{cosx}{sinx} = -cscx\ cotx$

33. (a) $x(t) = 4cos(\pi t/3)$, so $x(1) = 2$ ft.

       $v(t) = x'(t) = -(4\pi/3)sin(\pi t/3)$, so $v(1) = -2\pi\sqrt{3}/3$ ft/sec.

    (b) & (c) The graph of $v(t)$ is a sine curve which starts downward from
            the origin (when $t = 0$). It first reaches a maximum when
            $\pi t/3 = 3\pi/2$, or $t = 4.5$ seconds. The maximum value of v is
            $v(4.5) = 4\pi/3$ ft.

36. $f'(x) = \pi cos(\pi x/3) + (\pi/3)sin(\pi x/6)$, so $f'(1) = 2\pi/3$.

39. The limit is $f'(2)$ for $f(x) = sinx$.

    $f'(x) = cosx$, so $f'(2) = cos2 \approx -0.41615$

3. $\left[\frac{1}{3}\sin 3x\right]_0^{\pi/6} = \frac{1}{3}$ [Let $u = 3x$, if needed.]

6. $\int_0^{\pi/2} \frac{1}{2}\sin 2x\,dx = \left[-\frac{1}{4}\cos 2x\right]_0^{\pi/2} = \frac{1}{2}$ [Let $u = 2x$, if needed.]

9. $-\frac{1}{4}\cos^4 t + C$ [Let $u = \cos t$, if needed.]

12. $-\frac{1}{3}\sin^3(1/x) + C$ [Let $u = \sin(1/x)$, if needed.]

15. $\left[-\frac{1}{2}(\tan 2t)^{-1}\right]_{\pi/8}^{\pi/6} = -\frac{1}{2}(\sqrt{3}/3 - 1) \approx 0.2113$ [Let $u = \tan 2t$, if needed.]

18. $-\frac{1}{2}(\sin u)^{-2} + C = -\frac{1}{2}\csc^2 u + C$ [Let $z = \sin u$, if needed.]

21. $\int_{\pi/6}^{\pi/2} (\cos x)(\sin x)^{-1}dx = \left[\ln|\sin x|\right]_{\pi/6}^{\pi/2} = \ln 2$ [Let $u = \sin x$, if needed.]

24. $\left[\ln(1+\tan t)\right]_0^{\pi/4} = \ln 2$ [Let $u = 1+\tan t$, if needed.]

27. $2\int_0^{\pi/2} \pi[(\cos x)^{1/2}]^2 dx = 2\pi\left[\sin x\right]_0^{\pi/2} = 2\pi$

30. $-\frac{1}{2}\cos(x^2)+C$ [Let $u = x^2$, if needed.]

33. $\frac{1}{2}x^2\sin 2x - \int x\sin 2x\,dx$ [Letting $u = x^2$ and $dv = \cos 2x\,dx$.]

$= \frac{1}{2}x^2\sin 2x - [-\frac{1}{2}x\cos 2x + \int \frac{1}{2}\cos 2x\,dx]$ [Letting $u = x$ & $dv = \sin 2x\,dx$.]

$= \frac{1}{2}x^2\sin 2x + \frac{1}{2}x\cos 2x - \frac{1}{4}\sin 2x + C$

36. $\left[-\frac{1}{3}(2+\sin^2 t)\cos t\right]_{\pi/3}^{\pi/2} = \frac{11}{24}$ [Formula 24, with $t = u$]

39. (a) $-sin(1/x) + C$ [Let u = 1/x, if needed.]

(b) $sin(\ell nx) + C$ [Let u = ℓnx, if needed.]

(c) $\left[\frac{1}{8}sin^4(x^2)\right]_0^{(\pi/3)^{1/2}} = \frac{9}{128}$ [Let u = $sin(x^2)$.]

(d) $\left[-\frac{1}{4}xcos4x\right]_0^{\pi/4} + \frac{1}{4}\int_0^{\pi/4}cos4x\ dx = \frac{\pi}{16}$ [Letting u = x & dv = $sin4x$ dx.]

(e) $-\frac{1}{2}e^{cos^2x} + C$ [Let u = cos^2x, if needed.]

(f) $\left[-2cos(x^{1/2})\right]_{\pi^2/9}^{\pi^2/4} = 1$ [Let u = $x^{1/2}$, if needed.]

42. Let P(t) be the wolf population t months after 01/01/90.
    P'(t) = $450cos(\pi t/12)$ is the given rate of growth.
    Therefore, P(t) = $(5400/\pi)sin(\pi t/12)$ + C, for some C.
    P(0) = 6400 $\Rightarrow$ C = 6400, so P(t) = $(5400/\pi)sin(\pi t/12)$ + 6400.
    Thus, P(18) = $-5400/\pi$ + 6400 $\approx$ 4681 wolves on 07/01/91.

Problem Set 12.5 Applications of the Trigonometric Functions

3. $\left[\frac{8}{3}sin(\frac{3}{2}t)\right]_0^{\pi} = -\frac{8}{3}$

6. $\left[(480/\pi)sin[\pi(t-2)/4]\right]_0^6 = 480/\pi$

9. Amplitude is 4.
   Let $0 \le 3(t-2) \le 2\pi$.   Then $0 \le t-2 \le 2\pi/3$, so the period is $2\pi/3$.

12.

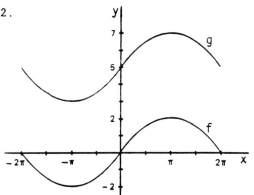

15. Period is $2\pi/B = 6$, so $B = \pi/3$.

   $A = \frac{1}{2}(30-10) = 10$.

   $A+d = 30$; $10+d = 30$, so $d = 20$.

18. (a) The period is $2\pi/(\pi/12) = 24$ months.

   (b) The graph is a sine curve starting upward from the origin (when $t = 0$). It first reaches its minimum value when $\pi t/12 = 3\pi/2$; i.e., when $t = 18$ months (July 1, 1981).

   (c) $P'(t) = 175\pi cos(\pi t/12)$, so $P'(3) = 87.5\pi\sqrt{2} \approx 389$ foxes per month.

   (d) $\int_0^{12} 2100sin(\pi t/12)\ dt = \left[(-25200/\pi)cos(\pi t/12)\right]_0^{12} = 50400/\pi$

   (Note that the constant term was omitted. It now needs to be put back into consideration.)
   Therefore, the average population is $8000 + \dfrac{50400/\pi}{12} \approx 9337$ foxes.

21. (a) The graph is a sine curve starting upward from (81,720) (when $t = 81$). It reaches its highest point when $(\pi/182)(t-81) = \pi/2$; i.e., when $t = 172$ days, so the "longest day" is June 21.

   (b) $L'(t) = (240\pi/182)cos[(\pi(t-81)/182]$, so $L'(81) = 240\pi/182 \approx 4.14$ minutes per day.

   (c) $\int_0^{81} 240sin[(\pi(t-81)/182]dt = \left[(-43680/\pi)cos[\pi(t-81)/182]\right]_0^{81} \approx -142$

   (Note that the constant term was omitted. It now needs to be put back into consideration.)
   Thus, average daylight is $720 - 142 = 578$ minutes (9 hr. 38 min.)

24. $20 = f(0) = A+B(0)$, so $A = 20$. Then $65 = f(6) = 20+B(1)$, so $B = 45$.

27. The graph appears in the answer section of the text.

3. $20\pi/3$ radians $= 20(60°) = 1200°$

6. $102.3° = 102.3\pi/180$ radians. Then $s = (15)(102.3\pi/180) \approx 26.78$ cm,

   and $A = \frac{1}{2}(15)^2(102.3\pi/180) \approx 200.87$ cm^2

9. $20sin76° \approx 19.4$ feet

12. (a) $\frac{1}{2}$ (b) -1   (c) -0.005622   (d) 2.1369

15. The graphs appear in the answer section of the text.

18. (a) $3(cos2x)(2) = 6cos2x$

    (b) $4[-cos(1/x)](-x^{-2}) = 4x^{-2}cos(1/x)$

    (c) $\frac{3}{2}(3x+sinx)^{1/2}(3+cosx)$

    (d) $(cos\sqrt{x^3+4})[\frac{1}{2}(x^3+4)^{-1/2}](3x^2) = 1.5x^2(x^3+4)^{-1/2} cos\sqrt{x^3+4}$

    (e) $\frac{1}{2}[sin(x^3+4)]^{-1/2}[cos(x^3+4)](3x^2) = 1.5x^2[sin(x^3+4)]^{-1/2}cos(x^3+4)$

    (f) $\dfrac{[cos(x^3)](2x) - (x^2)[-3x^2sin(x^3)]}{cos^2(x^3)} = 2xsec(x^3) + 3x^4sec(x^3)tan(x^3)$

21. (a) $-\frac{1}{3}sin(3x+1) + C$ [Let $u = 3x+1$, if needed.]

    (b) $\frac{1}{4}tan(4x) + C$ [Let $u = 4x$, if needed.]

    (c) $\ell n(1+sinx) + C$ [Let $u = 1+sinx$, if needed.]

    (d) $-\frac{2}{25}cos^{5/2}(5x) + C$ [Let $u = cos(5x)$, if needed.]

24. (a) $\left[\frac{1}{5}sin^5t\right]_0^{\pi/6} = \frac{1}{160}$

    (b) $\displaystyle\int_0^{\pi/6}(1-sin^2t)^{1/2}dt = \int_0^{\pi/6} cost \ dt = \left[sint\right]_0^{\pi/6} = \frac{1}{2}$

27. (a) $P(t) = 1800sin(\pi t/18) + C$, for some C.
$P(0) = 6400 \Rightarrow C = 6400$, so $P(t) = 1800sin(\pi t/18) + 6400$.

(b) Minimum value of C is 6400 - 1800 = 4600.

(c) $\int_0^6 1800sin(\pi t/18)\, dt = 16200/\pi$. Therefore, the average population
is $6400 + \dfrac{16200/\pi}{6} \approx 7259$.

30. $\int_0^{\pi^{1/2}} 4xsin(x^2)\, dx = \left[-2cos(x^2)\right]_0^{\pi^{1/2}} = 4$ [Let $u = x^2$, if needed.]

33. $\left[xsinx\right]_0^1 + \int_0^1 -sinx\, dx + \left[\frac{1}{2}sin(x^2)\right]_0^1$ [Let $u = x$ & $dv = cosx\, dx$ for the 1st term.]
[Let $u = x^2$, if needed, for the 2nd term.]

$= sin1 + (cos1 - 1) + \frac{1}{2}sin1$

$= \frac{3}{2}sin1 + cos1 - 1$

36. $2(x+2y)(1+2y') + 3(cos2y)(2y') = 3y'+2x$
Letting $x = 2$ & $y = 0$, obtain $y' = 0$, so the slope is 0.

39. (a) $\dfrac{dy}{dt} = -6sin2x\, \dfrac{dx}{dt} = (-6sin2x)(4) = -24sin2x$

When $x = \pi/12$, $\dfrac{dy}{dt} = -24(\frac{1}{2}) = -12$.

(b) $A(x) = \int_0^x 3cos2x\, dx$

$\dfrac{dA}{dt} = A'(x)\dfrac{dx}{dt} = (3cos2x)(4) = 12cos2x$

When $x = \pi/12$, $\dfrac{dA}{dt} = 19(\sqrt{3}/2) = 6\sqrt{3}$.

42. (a) $y' = 2\sqrt{3}cos2x - 2sin2x = 0$, if $2\sqrt{3}cos2x = 2sin2x$
$\sqrt{3} = tan2x$
$2x = \pi/3$
$x = \pi/6$

$y = 1$ at the left end point, $y = \sqrt{3}$ at the right end point, and
$y = 2$ when x is $\pi/6$. Therefore, the global maximum occurs at the
point $(\pi/6,2)$ and the global minimum occurs at the point $(0,1)$.

(b) $\int_0^{\pi/4}(\sqrt{3}sin2x + cos2x)dx = (1+\sqrt{3})/2$

Problem Set 13.1   First Order Equations

3. $y^{-1}dy = -4dt$, so $ln|y| = -4t+C$

$$y = \pm e^{-4t+C} = \pm e^{-4t}e^C = Ke^{-4t}, \ K \neq 0$$

6. $y^{-3}dy = x^2dx$, so $-\frac{1}{2}y^{-2} = -\frac{1}{3}x^3+C$; $y = [3/(K-2x^3)]^{1/2}$

9. $e^{-y}dy = e^{2x}dx$, so $-e^{-y} = \frac{1}{2}e^{2x}+C$; $y = -ln(K-\frac{1}{2}e^{2x})$

12. $tany \ dy = x^{-1}dx$, so $ln|secy| = ln|x| + C$; $secy = Kx$

15. $y^{-1}dy = x^{-2}dx$, so $ln|y| = -x^{-1}+C$; $y = e^{3-1/x}$

18. $e^ydy = (t+1)dt$, so $e^y = \frac{1}{2}t^2+t+C$

$y(0) = 2 \ \Rightarrow \ e^2 = C$, so $e^y = \frac{1}{2}t^2+t+e^2$; $y = ln(\frac{1}{2}t^2+t+e^2)$

NOTES: (1) In the following problems I.F. denotes an integrating factor.
(2) After multiplying each side of $y' + f(x)y = g(x)$ by I.F., the
equation simplifies to $\frac{d}{dx}[(I.F.)y] = (I.F.)f(x)$

21. $\frac{dy}{dx} + \frac{3}{x}y = 4$. $[I.F. = e^{\int 3/x \ dx} = x^3]$

$\frac{d}{dx}(x^3y) = 4x^3$, so $x^3y = x^4+C$; $y = x+Cx^{-3}$

24. $I.F. = e^{\int 1dx} = e^x$

$\frac{d}{dx}(e^xy) = e^{4x}$, so $e^xy = \frac{1}{4}e^{4x}+C$; $y = \frac{1}{4}e^{3x}+Ce^{-x}$

27. $I.F. = e^{\int 1/x \ dx} = x$

$\frac{d}{dx}(xy) = x^3+x$, so $xy = \frac{1}{4}x^4+ \frac{1}{2}x^2+C$

$y(2) = 5 \ \Rightarrow \ C = 4$, so $xy = \frac{1}{4}x^4+ \frac{1}{2}x^2+4$; $y = \frac{1}{4}x^3+ \frac{1}{2}x+4$.

30. I.F. $= e^{\int 5/x \ dx} = x^5$

$\frac{d}{dx}(x^5y) = 2x^{-1}$, so $x^5y = 2\ell nx+C$

$y(1) = 4 \Rightarrow C = 4$, so $x^5y = 2\ell nx+4$; $y = 2x^{-5}(2+\ell nx)$

33. $ydy = 5t^{3/2}dt$, so $\frac{1}{2}y^2 = 2t^{5/2}+C$; $y^2 = 4t^{5/2}+K$

36. I.F. $= e^{\int 2x/(1+x^2) \ dx} = 1+x^2$

$\frac{d}{dx}[(1+x^2)y] = 4x(1+x^2)^{1/2}$, so $y(1+x^2) = \frac{4}{3}(1+x^2)^{3/2}+C$

$y(1) = 3 \Rightarrow C = 6-\frac{8}{3}\sqrt{2}$, so $y(1+x^2) = \frac{4}{3}(1+x^2)^{3/2}+6-\frac{8}{3}\sqrt{2}$

$y = \frac{2}{3}(1+x^2)^{-1}[2(1+x^2)^{3/2}+9-4\sqrt{2}]$

39. $\frac{dy}{dx} = 4y$, so $y^{-1}dy = 4dx$; $\ell ny = 4x+C$.

$y(1) = e \Rightarrow C = -3$, so $\ell ny = 4x-3$; $y = e^{4x-3}$.

Problem Set 13.2 Applications of First-Order Equations

3. (a) $\frac{dy}{dx} = 6x^2y^2$, $y(0) = 2$

   (b) $y^{-2}dy = 6x^2dx$, so $-y^{-1} = 2x^3+C$

   $y(0) = 2 \Rightarrow C = -\frac{1}{2}$, so $-y^{-1} = 2x^3-\frac{1}{2}$; $y = 2(1-4x^3)^{-1}$

6. $\frac{dT}{dt} = k(T-22)$, $T(0) = -18$, $T(20) = 0$

   $(T-22)^{-1}dT = kdt$, so $\ell n|T-22| = kt+C$.

   $T(0) = -18 \Rightarrow C = \ell n40$, so $\ell n|T-22| = kt+\ell n40$.

   $T(20) = 0 \Rightarrow k = 0.05\ell n(22/40)$, so $\ell n|T-22| = 0.05\ell n(22/40)t+\ell n40$.

9. (a) Let y be the amount of salt (kg) in the tank t minutes after the brine begins to flow into the tank.

   $\frac{dy}{dt} = (0.25 \text{ kg/}\ell)(4 \ \ell/\text{min}) - (\frac{y}{200-(5-4)t} \text{ kg/}\ell)(5 \ \ell/\text{min})$, $y(0) = 0$

   which simplifies to $\frac{dy}{dt} = 1 - [5/(200-t)]y$, $y(0) = 0$

9. (b) $\frac{dy}{dt}$ + [5/(200-t)]y = 1   [I.F. = $e^{\int 5/(200-t)\ dt}$ = $(200-t)^{-5}$]

$\frac{d}{dt}[(200-t)^{-5})y]$ = $(200-t)^{-5}$, so $(200-t)^{-5}y$ = $0.25(200-t)^{-4}$+C

y(0) = 0 ⇒ C = $-\frac{1}{4}(200)^{-4}$, so $(200-t)^{-5}y$ = $\frac{1}{4}(200-t)^{-4}-\frac{1}{4}(200)^{-4}$;

y = $\frac{1}{4}(200-t)-\frac{1}{4}(200)^{-4}(200-t)^{5}$.

(c) y = $\frac{1}{8}(200-t)$ ⇒ $\frac{1}{8}(200-t)$ = $\frac{1}{4}(200-t)-\frac{1}{4}(200)^{-4}(200-t)^{5}$

t = $200[1-(0.5)^{1/4}]$ ≈ 31.8 minutes

(d) Let y'(t) = $-0.25 + 1.25(200)^{-4}(200-t)^{4}$ = 0

Then t = $200[1-(0.2)^{1/4}]$ ≈ 66.3 minutes.   y'(t) > 0 if t < 66.3
and y'(t) < 0 if t > 66.3, so y is maximum at that time.

12. Let y be the amount of ozone in the room, y(0) = A.

$\frac{dy}{dt}$ = $-(y/2000)(300)$ = -0.15y

$\frac{dy}{dt}$ + 0.15y = 0   [Let I.F. = $e^{\int 0.15dt}$ = $e^{0.15t}$.]

$\frac{d}{dt}(e^{0.15t}y)$ = 0, so $e^{0.15t}y$ = C.

y(0) = A ⇒ C = A, so y = $Ae^{-0.15t}$.

10% will be left when y = 0.10A.

$0.10A$ = $Ae^{-0.15t}$ if t = $(20/3)\ell n10$ ≈ 15.35 minutes.

15. (a) $40\frac{dv}{dt}$ = -40(32)-3.2v, v(0) = 0, s(0) = 12000

$\frac{dv}{dt}$ + 0.08v = -32   [Let I.F. = $e^{\int 0.08dt}$ = $e^{0.08t}$.]

$\frac{d}{dt}(e^{0.08t}v)$ = $-32e^{0.08t}$, so $ve^{0.08t}$ = $-400e^{0.08t}$+C; v = $-400+Ce^{-0.08t}$

v(0) = 0 ⇒ C = 400, so v = $-400+400e^{-0.08t}$.

$\frac{ds}{dt}$ = v = $-400+400e^{-0.08t}$, so s = $-400t-5000e^{-0.08t}$+K.

s(0) = 12000 ⇒ K = 17000, so s(t) = $17000-400t-5000e^{-0.08t}$.

(b) Let s(t) = $17000-400t-5000e^{-0.08t}$ = 0.   Try a few values of t.
s(42) ≈ 26.3   and   s(43) ≈ -360.3, so it hits in about 42 seconds.

18. $\frac{dC}{dt}$ = kA(P-C), C(0) = 0

21. Let P be the population after t days.

$$\frac{dP}{dt} = 40t+30t^2, \quad P(0) = 6000$$

Therefore, $P(t) = 20t^2+10t^3+C$.
$P(0) = 6000 \Rightarrow C = 6000$, so $P(t) = 20t^2+10t^3+6000$.
Then $P(10) = 18000$ insects.

24. Let x be the amount of salt (lbs) in tank A, y be amount in tank B t minutes after the process begins.

Tank A: $\frac{dx}{dt} = - (x/100)(2) = -0.02x, \quad x(0) = 50$

$\frac{dx}{dt} + 0.02x = 0$ [Let I.F. $= e^{\int 0.02dt} = e^{0.02t}$.]

$\frac{d}{dt}(e^{0.02t}x) = 0$, so $e^{0.02t}x = C$ or $x = Ce^{-0.02t}$.

$x(0) = 50 \Rightarrow C = 50$, so $x = 50e^{-0.02t}$.

Tank B: $\frac{dy}{dt} = \frac{x}{50} - \frac{2y}{200}, \quad y(0) = 50$

$\frac{dy}{dt} = 0.02x - 0.01y = 0.02(50e^{-0.02t}) - 0.01y$

$\frac{dy}{dt} + 0.01y = e^{-0.02t}$ [Let I.F. $= e^{\int 0.01dt} = e^{0.01t}$.]

$\frac{d}{dt}(e^{0.01t}y) = e^{-0.01t}$, so $e^{0.01t}y = -100e^{-0.01t}+K$;

$$y = -100e^{-0.02t}+Ke^{-0.01t}$$

$y(0) = 50 \Rightarrow K = 150$, so $y(t) = 150e^{-0.01t}-100e^{-0.02t}$

Problem Set 13.3   Homogeneous Second-Order Linear Equations

3. $y' = \frac{1}{3}sin3t + C_1; \quad y = -\frac{1}{9}cos3t + C_1t + C_2$.

6. $y' = -e^{-t}-4t+C_1; \quad y = e^{-t}-2t^2 + C_1t + C_2$.

9. Roots of auxiliary equation, $r^2-4r = 0$, are 0 & 4, so the general solution is $y = C_1 + C_2e^{4x}$.

12. Roots of auxiliary equation, $r^2+8r+15 = 0$, are -5 & -3, so the general solution is $y = C_1e^{-5x} + C_2e^{-3x}$.

15. Roots of auxiliary equation, $r^2+r-1 = 0$, are $(-1 \pm \sqrt{5})/2$, so the general solution is $y = C_1e^{(-1-\sqrt{5})x/2} + C_2e^{(-1+\sqrt{5})x/2}$.

18. Roots of auxiliary equation, $4r^2-12r+7 = 0$, are $(3 \pm \sqrt{2})/2$, so the general solution is $y = C_1e^{(3-\sqrt{2})t/2} + C_2e^{(3+\sqrt{2})t/2}$.

21. Roots of auxiliary equation, $9r^2+6r+1 = 0$, are -1/3 & -1/3, so the general solution is $x = C_1e^{-t/3} + C_2te^{-t/3}$.

24. Roots of auxiliary equation, $r^2+49 = 0$, are -7i & 7i, so the general solution is $y = C_1 sin7t + C_2 cos7t$.

27. Roots of auxiliary equation, $r^2-10r+34 = 0$, are 5-3i & 5+3i, so the general solution is $y = e^{5x}(C_1 sin3x + C_2 cos3x)$.

30. Roots of auxiliary equation, $r^2+3r-4 = 0$, are -4 & 1, so the general solution is $y = C_1e^{-4x} + C_2e^{x}$. $y' = -4C_1e^{-4x} + C_2e^{x}$.
$y(0) = 8 \Rightarrow C_1 + C_2 = 8$ and $y'(0) = -7 \Rightarrow -4C_1+C_2 = -7$,
so $C_1 = 3$, $C_2 = 5$. Therefore, $y = 3e^{-4x} + 5e^{x}$.

33. Roots of auxiliary equation, $r^2+1 = 0$, are -i & i, so the general solution is $x = C_1 sint + C_2 cost$.
$x(0) = 0 \Rightarrow C_2 = 0$ and $x(\pi/2) = 3 \Rightarrow C_1 = 3$,
Therefore, $x = 3sint$.

36. Roots of auxiliary equation, $r^3+5r^2+6r = 0$, are -3, -2 & 0, so the general solution is $y = C_1e^{-3x} + C_2e^{-2x} + C_3$.

39. $\dfrac{d^2}{dt^2}[C_1u(t) + C_2v(t)] + a\dfrac{d}{dt}[C_1u(t) + C_2v(t)] + b[C_1u(t) + C_2v(t)]$

$= [C_1u''(t) + C_2v''(t)] + a[C_1u'(t) + C_2v'(t)] + b[C_1u(t) + C_2v(t)]$

$= C_1[u''(t) + au'(t) + bu(t)] + C_2[v''(t) + av'(t) + bv(t)]$

$= C_1(0) + C_2(0) = 0$

42. Roots of auxiliary equation, $r^3-4r^2 = 0$, are 0, 0 & 4, so the

general solution is $x = C_1 + C_2t + C_3e^{4t}$; $x' = C_2 + 4C_3e^{4t}$; $x'' = 16C_3e^{4t}$.

$\begin{array}{ll} x(0) = 1 \Rightarrow & C_1 + C_3 = 1, \\ x'(0) = -1 \Rightarrow & C_2 + 4C_3 = -1, \\ x''(0) = -16 \Rightarrow & 16C_3 = -16, \end{array}$

so $C_1 = 2$, $C_2 = 3$, $C_3 = -1$.   Therefore, $x = 2+3t-e^{4t}$.

Problem Set 13.4   Nonhomogeneous Second-Order Linear Equations

3. Roots of $r^2+4 = 0$ are $-2i$ & $2i$, so $y_h = C_1 sin2x + C_2 cos2x$.

Let $y_p = Ax+B$; $y_p' = A$; $y_p'' = 0$.

Then $0 + 4(Ax+B) = 4x+8$, so $A = 1$, $B = 2$.

Therefore, $y = C_1 sin2x + C_2 cos2x + x + 2$ is the general solution.

6. $y_h = C_1 sin3x + C_2 cos3x$ [See Problem 37(c), Section 13.3.]

Let $y_p = Asinx$; $y_p' = Acosx$; $y_p'' = -Asinx$.

Then $-Asinx + 9Asinx = 3sinx$, so $A = \frac{3}{8}$.

Therefore, $y = C_1 sin3x + C_2 cos3x + \frac{3}{8}sinx$ is the general solution.

9. Roots of $r^2-3r+2 = 0$ are 1 & 2, so $y_h = C_1 e^x + C_2 e^{2x}$.

Let $y_p = Ax^2+Bx+C$; $y_p' = 2Ax+B$; $y_p'' = 2A$.

Then $2A - 3(2Ax+B) + 2(Ax^2+Bx+C) = x^2-1$, so $A = \frac{1}{2}$, $B = \frac{3}{2}$, $C = \frac{5}{4}$.

Therefore, $y = C_1 e^x + C_2 e^{2x} + \frac{1}{2}x^2 + \frac{3}{2}x + \frac{5}{4}$ is the general solution.

12. Roots of $r^2-7r+10 = 0$ are 2 & 5, so $y_h = C_1e^{2x} + C_2e^{5x}$.

Let $y_p = A\sin x + B\cos x$; $y_p' = A\cos x - B\sin x$; $y_p'' = -A\sin x - B\cos x$.

∴ $(-A\sin x - B\cos x) - 7(A\cos x - B\sin x) + 10(A\sin x + B\cos x) = 130\cos x$,

so $A = -7$, $B = 9$.

Therefore, $y = C_1e^{2x} + C_2e^{5x} - 7\sin x + 9\cos x$ is the general solution.

15. Roots of $r^2-6r+8 = 0$ are 2 & 4, so $y_h = C_1e^{2x} + C_2e^{4x}$.

Let $y_p = Axe^{2x}$; $y_p' = Ae^{2x}(1+2x)$; $y_p'' = 4Ae^{2x}(1+x)$.

Thus, $[4Ae^{2x}(1+x)] - 6[Ae^{2x}(1+2x)] + 8[Axe^{2x}] = 4e^{2x}$, so $A = -2$.

Therefore, $y = C_1e^{2x} + C_2e^{4x} - 2xe^{2x}$.

18. $y_h = C_1\sin 3x + C_2\cos 3x$ [See Problem 6.]

Let $y_p = Ax\sin 3x + Bx\cos 3x$; $y_p' = (A-3Bx)\sin 3x + (3Ax+B)\cos 3x$;

$y_p'' = (-9Ax-6B)\sin 3x + (6A-9Bx)\cos 3x$.

Thus, $[(-9Ax-6B)\sin 3x + (6A-9Bx)\cos 3x]$
    $+ 9[(A-3Bx)\sin 3x + (3Ax+B)\cos 3x] + [Ax\sin 3x + Bx\cos 3x] = 12\cos 3x$.

$6A\cos 3x - 6B\sin 3x = 12\cos 3x$, so $A = 2$, $B = 0$.

Therefore, $y = C_1\sin 3x + C_2\cos 3x + 2x\sin 3x$.

21. $y = C_1\sin 2x + C_2\cos 2x + 2\cos x$; $y' = 2C_1\cos 2x - 2C_2\sin 2x - 2\sin x$.

$y(\pi/2) = 4 \Rightarrow C_2 = -4$ and $y'(\pi/2) = 6 \Rightarrow C_1 = -4$.

Therefore, $y = -4\sin 2x - 4\cos 2x + 2\cos x$

24. $y = C_1\sin 2x + C_2\cos 2x + x + 2$; $y' = 2C_1\cos 2x - 2C_2\sin 2x + 1$.

$y(0) = 0 \Rightarrow C_2 = -2$ and $y'(\pi/4) = 6 \Rightarrow C_1 = 4-\frac{\pi}{4}$.

Therefore, $y = (4-\frac{\pi}{4})\sin 2x - 2\cos 2x + x + 2$.

Problem Set 13.4   141

27. (a) $y" + \frac{q}{m}y' + \frac{k}{m}y = 0$, $y(0) = -1$, $y'(0) = 0$.

$y" + 0.05y' + 4y = 0$ [$q = 0.125$, $k = 10$, $m = 2.5$]

(b) Roots of $r^2 + 0.05r + 4 = 0$ are approximately $-0.025 \pm 2i$.

$y = e^{-0.025t}(C_1 sin2t + C_2 cos2t)$;

$y' = e^{-0.025t}(2C_1 cos2t - 2C_2 sin2t) - 0.025e^{-0.025t}(C_1 sin2t + C_2 cos2t)$

$y(0) = -1 \Rightarrow C_2 = -1$ and then $y'(0) = 0 \Rightarrow C_1 = -0.0125$.

Therefore, $y = e^{-0.025t}(-0.0125sin2t - cos2t)$.

Thus, $y \approx e^{-0.025t}(-cos2t)$.

(c) The graph appears in the answer section of the text.

30. $y_h = C_1 e^{-2x} + C_2 e^{2x}$ [from Problem 1]

Let $y_p = Asinx$; $y_p' = Acosx$; $y_p" = -Asinx$.

Then $-Asinx - 4Asinx = 5sinx$, so $A = -1$.

Therefore, $y = C_1 e^{-2x} + C_2 e^{2x} - sinx$; $y' = -2C_1 e^{-2x} + 2C_2 e^{2x} - cosx$.

$y(0) = 1 \Rightarrow 1 = C_1 + C_2$ and $y'(0) = 5 \Rightarrow 6 = -2C_1 + 2C_2$.

Then $C_1 = -1$, $C_2 = 2$.   Thus, $y = -e^{-2x} + 2e^{2x} - sinx$.

33. $y_h = C_1 sin2x + C_2 cos2x$ [From Problem 3]

Let $y_p = Axsin2x + Bxcos2x$; $y_p' = (A-2Bx)sin2x + (2Ax+B)cos2x$;

$y_p" = (-4Ax-4B)sin2x + (4A-4Bx)cos2x$.

$\therefore$ $[(-4Ax-4B)sin2x + (4A-4Bx)cos2x] + [Axsin2x + Bxcos2x] = 8sin2x$.

$4Acos2x - 4Bsin2x = 8sin2x$, so $A = 0$, $B = -2$.

Therefore, $y = C_1 sin2x + C_2 cos2x - 2xcos2x$;

$y' = 2C_1 cos2x - 2C_2 sin2x - 2cos2x + 4xsin2x$.

$y(0) = 9 \Rightarrow 9 = C_2$ and $y'(0) = 3 \Rightarrow C_1 = \frac{5}{2}$.

Then $y = \frac{5}{2}sin2x + 9cos2x - 2xcos2x$.

36. Roots of $r^3-5r^2+4r = 0$ are 0, 1, & 4, so $y_h = C_1 + C_2e^x + C_3e^{4x}$.

Let $y_p = Ae^{2x}$; $y'_p = 2Ae^{2x}$; $y''_p = 4Ae^{2x}$; $y_p^{(3)} = 8Ae^{2x}$.

Then $(8Ae^{2x}) - 5(4Ae^{2x}) + 4(2Ae^{2x}) = 8e^{2x}$, so $A = -2$.

Therefore, $y = C_1 + C_2e^x + C_3e^{4x} - 2e^{2x}$; $y' = C_2e^x + 4C_3e^{4x} - 4e^{2x}$;
$$y'' = C_2e^x + 16C_3e^{4x} - 8e^{2x}.$$

$\begin{aligned}
y(0) &= 9 \Rightarrow C_1 + C_2 + C_3 = 11, \\
y'(0) &= 3 \Rightarrow C_2 + 4C_3 = 7, \\
y''(0) &= -11 \Rightarrow C_2 + 16C_3 = -3,
\end{aligned}$

so $C_1 = \frac{3}{2}$, $C_2 = \frac{31}{3}$, $C_3 = -\frac{5}{6}$.

Therefore, $y = \frac{3}{2} + \frac{31}{3}e^x - 2e^{2x} - \frac{5}{6}e^{4x}$.

39. This is Problem 18 with t instead of x, so we can go right to

$y = C_1 sin3t + C_2 cos3t + 2tsin3t$;

Then $y' = 3C_1 cos3t - 3C_2 sin3t + 2sin3t + 6tcos3t$.

$y(0) = 1 \Rightarrow 1 = C_2$ and $y'(0) = 0 \Rightarrow C_1 = 0$, so $y = cos3t + 2tsin3t$

Chapter 13 Review Problem Set

3. $y^{-1}dy = x(x^2+1)^{-1}dx$

$ln|y| = \frac{1}{2}ln(x^2+1) + C$; $y = K(x^2+1)^{1/2}$

6. $\frac{d}{dx}(x^2y) = 4x+3x^3$ [Using I.F. $= e^{\int 2/x \, dx} = x^2$]

$x^2y = 2x^2+ \frac{3}{4}x^4 + C$; $y = 2+ \frac{3}{4}x^2+Cx^{-2}$.

9. Roots of auxiliary equation, $r^2-16 = 0$, are -4 & 4, so the general

solution is $y = C_1e^{-4x} + C_2e^{4x}$.

12. Roots of auxiliary equation, $r^2+25 = 0$, are -5i & 5i, so the general

solution is $y = C_1 sin5t + C_2 cos5t$.

15. Roots of auxiliary equation, $r^2+4r-1 = 0$, are $-2 \pm \sqrt{5}$, so the general solution is $y = C_1 e^{(-2-\sqrt{5})x} + C_2 e^{(-2+\sqrt{5})x}$.

18. Roots of auxiliary equation, $r^2-8r+16 = 0$, are 4 & 4, so the general solution is $y = C_1 e^{4t} + C_2 t e^{4t}$.

21. $y_h = C_1 e^{-4x} + C_2 e^{4x}$ [From Problem 9]

Let $y_p = A\sin x + B\cos x$; $y_p' = A\cos x - B\sin x$; $y_p'' = -A\sin x - B\cos x$.

$\therefore$ $(-A\sin x - B\cos x) - 16(A\sin x + B\cos x) = \cos x$, so $A = 0$, $B = -\frac{1}{17}$.

Therefore, $y = C_1 e^{-4x} + C_2 e^{4x} - \frac{1}{17}\cos x$ is the general solution.

24. Roots of $r^2-4r-5 = 0$ are -1 & 5, so $x_h = C_1 e^{-t} + C_2 e^{5t}$.

Let $x_p = Ae^t+Bt+C$; $x_p' = Ae^t+B$; $x_p'' = Ae^t$.

Then $(Ae^t) - 4(Ae^t+B) - 5(Ae^t+Bt+C) = 16e^t+5t$

$-8Ae^t - 5Bt + (-4B-5C) = 16e^t+5t$, so $A = -2$, $B = -1$, $C = \frac{4}{5}$.

Thus, $y = C_1 e^{-t} + C_2 e^{5t} - 2e^t - t + \frac{4}{5}$.

27. $\frac{dy}{dx} = \frac{y^2}{x^2}$ or $y^{-2}dy = x^{-2}dx$.

$-y^{-1} = -x^{-1} + C$ or $y^{-1} = x^{-1} - C$

At $(\frac{1}{2}, -\frac{1}{4})$, $-4 = 2-C$, so $C = 6$.

Therefore, $y^{-1} = x^{-1} - 6$; $y = x(1-6x)^{-1}$

30. $y = C_1 e^x + C_2 e^{3x}$; $y' = C_1 e^x + 3C_2 e^{3x}$

$y(0) = 3 \Rightarrow 3 = C_1 + C_2$,
$y'(0) = 7 \Rightarrow 7 = C_1 + 3C_2$, so $C_1 = 1$, $C_2 = 2$.

Then $y = e^x + 2e^{3x}$.

33. (a) $\frac{dT}{dt} = k(T-75)$, $T(0) = 450$, $T(20) = 200$.

$(T-75)^{-1}dT = kdt$, so $\ln(T-75) = kt + C$.

$T(0) = 450 \Rightarrow C = \ln(375)$, so $\ln(T-75) = kt + \ln(375)$

$T(20) = 200 \Rightarrow k = -0.05\ln3$, so $\ln(T-75) = (-0.05\ln3)t + \ln(375)$.

Therefore, $T = 75 + 375e^{(-0.05\ln3)t}$.

(b) It $T = 120$, then $t = [-20\ln(45/375)]/(\ln3) \approx 38.6$ minutes.

36. (a) Let y be the amount of salt (kg) in the vat t minutes after the process begins.

$\frac{dy}{dt} = (-y/1500)(5)$, $y(0) = 40$

$\frac{dy}{dt} + \frac{1}{300}y = 0$ $\quad [\text{I.F.} = e^{\int 1/300\ dt} = e^{t/300}]$

$\frac{d}{dt}(e^{t/300}y) = 0$, so $e^{t/300}y = C$ or $y = Ce^{-t/300}$.

$y(0) = 40 \Rightarrow C = 40$, so $y(t) = 40e^{-t/300}$.

Therefore, $y(30) = 40e^{-0.1} \approx 36.2$ kg.

(b) $y = (0.01)(1500) = 15$ kg is present in a 0.01 concentration.

Then $t = 300\ln(8/3) \approx 294$ minutes (almost 5 hours).

39. $a(t) = -32$; $v(t) = -32t+v_0 = -32t-32$;

$s(t) = -16t^2-32t+s_0 = -16t^2-32t+400$.

$s = 0$, if $t = -1 + \sqrt{26} \approx 4.099$; and $v(-1 + \sqrt{26}) \approx -163.169$.

It will hit the ground in about 4.1 seconds and will be traveling at about 163 feet per second.

42. Roots of $r^2-5r+4 = 0$, are 1 & 4, so $y_h = C_1e^x + C_2e^{4x}$.

$y_p = Ax^3 + Bx^2 + Cx + D + Exe^x + Fe^{2x}$.

45. (a) Let $y = x^r$, $y' = rx^{r-1}$, $y'' = r(r-1)x^{r-2}$.

Let $6x^2[r(r-1)x^{r-2}] + 7x[rx^{r-1}] - 2[x^r] = 0$, and solve for r.

$x^r(3r+2)(2r-1) = 0$, so $r = -\frac{2}{3}$ or $r = \frac{1}{2}$.

(b) $y = x^{-2/3}$ and $y = x^{1/2}$ are independent solutions, so the general

solution is $y = C_1 x^{-2/3} + C_2 x^{1/2}$.

48. $\left[\dfrac{1}{y(y-1)(y-3)}\right] dy = 6dt$

$\left[\dfrac{1/3}{y} - \dfrac{1/2}{y-1} + \dfrac{1/6}{y-3}\right] dy = 6dt$; $\left[\dfrac{2}{y} - \dfrac{3}{y-1} + \dfrac{1}{y-3}\right] dy = 12dt$

$\ln y^2 - \ln(y-1)^3 + \ln(y-3) = 36t + C$

$\ln[y^2(y-3)(y-1)^{-3}] = 36t + C$

$y(0) = 4 \Rightarrow C = \ln(16/27)$, so $\ln[y^2(y-3)(y-1)^{-3}] = 36t + \ln(16/27)$.

$y^2(y-3)(y-1)^{-3} = \frac{16}{27}e^{36t}$

CHAPTER 14   NUMERICAL CALCULUS: APPROXIMATION

Problem Set 14.1   Numerical Integration

3. (a) $\left[-cosx\right]_0^3 = -cos3 + 1 \approx 1.98999$

$f(x) = sinx$, $h = \dfrac{3-0}{6} = 0.5$, partition is $\{0, 0.5, 1, 1.5, 2, 2.5, 3\}$.

(b) $0.5[sin(0.5)+sin(1)+sin(1.5)+sin(2)+sin(2.5)+sin(3)] \approx 1.98364$

(c) $\dfrac{0.5}{2}[sin(0)+2sin(0.5)+2sin(1)+2sin(1.5)+2sin(2)+2sin(2.5)+sin(3)]$
≈ 1.98364

(d) $\dfrac{0.5}{3}[sin(0)+4sin(0.5)+2sin(1)+4sin(1.5)+2sin(2)+4sin(2.5)+sin(3)]$
≈ 1.99070

6. (a) $\left[x\ell nx-x\right]_1^4 = (4\ell n4-4) - (-1) \approx 2.54518$

$f(x) = \ell nx$, $h = \dfrac{4-1}{6} = 0.5$, partition is $\{1, 1.5, 2, 2.5, 3, 3.5, 4\}$.

(b) $0.5[\ell n(1.5)+\ell n(2)+\ell n(2.5)+\ell n(3)+\ell n(3.5)+\ell n(4)] \approx 2.87629$

(c) $\dfrac{0.5}{2}[\ell n(1)+2\ell n(1.5)+2\ell n(2)+2\ell n(2.5)+2\ell n(3)+2\ell n(3.5)+\ell n(4)]$
≈ 2.52971

(d) $\dfrac{0.5}{3}[\ell n(1)+4\ell n(1.5)+2\ell n(2)+4\ell n(2.5)+2\ell n(3)+4\ell n(3.5)+\ell n(4)]$
≈ 2.54465

9. (a) $f(x) = sin(x^2)$, $h = \dfrac{2-0}{4} = 0.5$, partition is $\{0, 0.5, 1, 1.5, 2\}$.

$\dfrac{0.5}{3}[sin(0)+4sin(0.25)+2sin(1)+4sin(2.25)+sin(4)] \approx 0.83801$

(b) $h = \dfrac{2-0}{8} = 0.25$, partition is $\{0, 0.25, 0.5, 0.75, 1, 1.25, 1.5, 1.75, 2\}$.

$\dfrac{0.25}{3}[sin(0)+4sin(0.0625)+2sin(0.25)+4sin(0.5625)+2sin(1)$
$+4sin(1.5625)+2sin(2.25)+4sin(3.0625)+sin(4)] \approx 0.80634$

(c) $h = \dfrac{2-0}{16} = 0.125$, partition is $\{0, 0.125, 0.25, 0.375, 0.5, 0.625, 0.75,$
$0.875, 1, 1.125, 1.25, 1.375, 1.5, 1.625, 1.75, 1.875, 2\}$.

$\dfrac{0.125}{3}[sin(0)+4sin(0.015625)+2sin(0.0625)+4sin(0.140625)+2sin(0.25)$
$+4sin(0.390625)+2sin(0.5625)+4sin(0.765625)+2sin(1)$
$+4sin(1.265625)+2sin(1.5625)+4sin(1.890625)+2sin(2.25)$
$+4sin(2.640625)+2sin(3.0625)+4sin(3.515625)+sin(4)]$
≈ 0.80135

12. (a) $f(x) = \ell nx$, $f'(x) = x^{-1}$, $f''(x) = -x^{-2}$.

$|f''(x)| = |x|^{-2} \le 1$ for x in [1,4].

Therefore, $|E| \le \dfrac{(4-1)^3}{12(6)^2}(1) = 0.0625$.

(b) $f^{(3)}(x) = 2x^{-3}$, $f^{(4)}(x) = -6x^{-4}$.

$|f^{(4)}(x)| = 6|x|^{-4} \le 6$ for x in [1,4].

Therefore, $|E| \le \dfrac{(4-1)^5}{180(6)^4}(6) = 0.00625$.

15. $\dfrac{20}{3}[(40) + 4(60) + 2(65) + 4(70) + 2(90) + 4(100) + 90] \approx 9067$ ft^2.

18. $f(x) = \sqrt{\ell nx}$, $h = \dfrac{2-1}{10} = 0.1$, partition is {0,0.1,0.2,0.3,0.4,0.5,0.6, 0.7,0.8,0.9,1}.

$\dfrac{0.1}{2}[0+2\sqrt{\ell n1.1}+2\sqrt{\ell n1.2}+2\sqrt{\ell n1.3}+2\sqrt{\ell n1.4}+2\sqrt{\ell n1.5}+2\sqrt{\ell n1.6}+2\sqrt{\ell n1.7}+2\sqrt{\ell n1.8}$
$+2\sqrt{\ell n1.9}+\sqrt{\ell n2}] \approx 0.58629$.

21. The integral is the area of the first quadrant portion of the region bounded by the unit circle centered at the origin. Its area is $\pi/4$.

h = 0.1, partition is {0,0.1,0.2,0.3,0.4,0.5,0.6,0.7,0.8,0.9,1}.

The integral is approximately

$\dfrac{0.1}{3}[1+4\sqrt{0.99}+2\sqrt{0.96}+4\sqrt{0.91}+2\sqrt{0.84}+4\sqrt{0.75}+2\sqrt{0.64}+4\sqrt{0.51}+2\sqrt{0.36}+4\sqrt{0.19}]$
≈ 0.781752039

Therefore, $\pi \approx 4(0.781752039) \approx 3.12701$.

Problem Set 14.2 Approximating Functions by Polynomials

3. $f(x) = (x+2)^{1/2}$, $f'(x) = \frac{1}{2}(x+2)^{-1/2}$, $f''(x) = -\frac{1}{4}(x+2)^{-3/2}$.

$f(2) = 2$, $f'(2) = \frac{1}{4}$, $f''(2) = -\frac{1}{32}$.

(a) $P_1(x) = 2 + \frac{1}{4}(x-2)$

(b) $P_2(x) = 2 + \frac{1}{4}(x-2) - \frac{1}{64}(x-2)^2$

(c) $P_1(2.01) = 2.0025$, $P_2(2.01) \approx 2.00249844$, $f(2.01) \approx 2.00249844$.

6. $f(x) = x(x^2+3)^{-1}$, $f'(x) = (x^2+3)^{-2}(3-x^2)$, $f''(x) = (x^2+3)^{-3}(2x^3-18x)$.

$f(2) = \frac{2}{7}$, $f'(2) = -\frac{1}{49}$, $f''(2) = -\frac{20}{343}$.

(a) $P_1(x) = \frac{2}{7} - \frac{1}{49}(x-1)$

(b) $P_2(x) = \frac{2}{7} - \frac{1}{49}(x-1) - \frac{10}{343}(x-1)^2$

(c) $P_1(2.01) \approx 0.02855102$, $P_2(2.01) \approx 0.28550729$, $f(2.01) \approx 0.28550731$

9. $f(x) = e^{-x}$, $f'(x) = -e^{-x}$, $f''(x) = e^{-x}$, $\ldots$, $f^{(n)}(x) = (-1)^n e^{-x}$.

$f(0) = 1$, $f'(0) = -1$, $f''(0) = 1$, $\ldots$, $f^{(n)}(0) = (-1)^n$

$P_n(x) = 1 - x + \frac{x^2}{2!} - \frac{x^3}{3!} + \cdots + (-1)^n \frac{x^n}{n!}$

12. $f(x) = \ln(1+x)$, $f'(x) = (1+x)^{-1}$, $f''(x) = -(1+x)^{-2}$, $f^{(3)}(x) = 2(1+x)^{-3}$,

$f^{(4)}(x) = -3!(1+x)^{-4}$, $\ldots$, $f^{(n)}(x) = (-1)^{n-1}(n-1)!(1+x)^{-n}$.

$f(0) = 0$, $f'(0) = 1$, $f''(0) = -1$, $f^{(3)}(0) = 2$, $f^{(4)}(0) = -3!$, $\ldots$,

$f^{(n)}(0) = (-1)^{n-1}(n-1)!$.

$P_n(x) = 0 + x - \frac{x^2}{2} + \frac{x^3}{3} + \cdots + (-1)^{n-1} \frac{x^n}{n}$

15. $|f^{(4)}(x)| = e^{-x} < e^{0.5}$ on $[-0.5, 0.5]$.

$|E(x)| \leq \frac{(0.5)^4}{4!}(e^{0.5}) < 0.00430$

18. $|f^{(5)}(x)| = 6[(x+1)^{-4}+4(x+1)^{-5}] \leq 6[(2)^{-4}+4(2)^{-5}] = 0.1875$ on $[1, 1.1]$.

$|E(x)| \leq \frac{(0.1)^5}{5!}(0.1875) = 0.00000009375$

21. $|f^{(6)}(x)| = sin x < 1$ on $[\pi/2, 1.68]$.

$|E(x)| \leq \frac{(1.68 - \pi/2)}{6!}(1) < 0.00000000236$

24. Let $f(x) = 2(x-2)^3 + (x-2)^2 - 4(x-2) + 5$, $\quad f(-3) = -200$
$\quad f'(x) = 6(x-2)^2 + 2(x-2) - 4$ $\qquad\qquad\qquad f'(-3) = 136$
$\quad f''(x) = 12(x-2)$ $\qquad\qquad\qquad\qquad\qquad f''(-3) = -58$
$\quad f^{(3)}(x) = 12$ $\qquad\qquad\qquad\qquad\qquad\quad f^{(3)}(-3) = 12$

$f^{(4)}(x) = 0$, so $|E(x)| \leq 0$; i.e., $E(x) = 0$, so the Taylor polynomial for $f(x)$ based at $x = -3$ represents $f(x)$.

$$P_3(x) = -200 + 136(x+3) - \frac{58}{2!}(x+3)^2 + \frac{12}{3!}(x+3)^3,$$

so $a = 2$, $b = -29$, $c = 136$, $d = -200$.

27. $f(x) = (\cos x)^{1/2}$, $\quad f'(x) = -\frac{1}{2}(\cos x)^{-1/2}\sin x$,

$\quad f''(x) = -\frac{1}{4}(\cos x)^{-3/2}\sin^2 x - \frac{1}{2}(\cos x)^{1/2}$.

$f(0) = 1$, $f'(0) = 0$, $f''(0) = -\frac{1}{2}$, so $P_2(x) = 1 - \frac{1}{4}x^2$.

Therefore, $\displaystyle\int_0^1 \sqrt{\cos x}\ dx \approx \int_0^1 (1 - \frac{1}{4}x^2)\,dx = \frac{11}{12}$.

Problem Set 14.3   Solving Equations Numerically

3. Let $f(x) = x^4 + 4x^3 + 2 = 0$; $\quad f'(x) = 4x^3 + 12x^2$.

$$x_{n+1} = x_n - \frac{x_n^4 + 4x_n^3 + 2}{4x_n^3 + 12x_n^2}$$

n	x_n
1	-4
2	-3.96875
3	-3.96798798
4	-3.96798753

Root ≈ -3.9680.

6. Let $f(x) = x^3 - 3x - 10 = 0$; $\quad f'(x) = 3x^2 - 3$. Note that f is increasing on $(-\infty, -1)$ and on $(1, \infty)$. The only root is about halfway between 2 & 3.

$$x_{n+1} = x_n - \frac{x_n^3 - 3x_n - 10}{3x_n^2 - 3}$$

n	x_n
1	2.5
2	2.619047619
3	2.612904811
4	2.612887865
5	2.612887865

Root ≈ 2.612888.

9. See graph in answer section of text. Let $f(x) = 2x - cos3x = 0$. Then $f'(x) = 2 + 3sin3x$. Let $x_1 = 1/3$.

$$x_{n+1} = x_n - \frac{2x_n + cos3x_n}{2 + 3sin3x_n}$$

n	x_n
1	1/3
2	0.305403884
3	0.304952287
4	0.304952159
5	0.304952159

Root ≈ 0.304952.

12. Let $f(x) = x - 2 + \ell nx = 0$; $f'(x) = 1 + x^{-1}$. Let $x_1 = 1.6$.

$$x_{n+1} = x_n - \frac{x_n - 2 + \ell nx_n}{1 + 1/x_n}$$

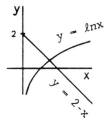

n	x_n
1	1.6
2	1.556920844
3	1.557145593
4	1.557145599

Root ≈ 1.557146

15. Let $f(x) = x^3 - 4x + 1 = 0$; $f'(x) = 3x^2 - 4$. Then f is increasing on $(-\infty, -\sqrt{4/3})$ and on $(\sqrt{4/3}, \infty)$. $f(-3) = -14$ and $f(-2) = 1$, so there is a root between -3 and -2; $f(0) = 1$ and $f(1) = -2$, so there is a root between 0 and 1; $f(1) = -2$ and $f(2) = 1$, so there is a root between 1 and 2.

Let $x_{n+1} = x_n - \dfrac{x_n^3 - 4x_n + 1}{3x_n^2 - 4}$

n	x_n
1	-2.1
2	-2.11505958
3	-2.11490755
4	-2.11490754

n	x_n
1	0.3
2	0.253619302
3	0.254101641
4	0.254101688

n	x_n
1	1.7
2	1.88993576
3	1.861518585
4	1.860806297
5	1.860805853
6	1.860805853

Roots are approximately -2.11491, 0.25410, 1.86081.

18. The graph of $y = 5\cos x$ will cross the graph of $y = x$ near (to the left of) where x is $-\pi/2$ and where x is $-\pi$.

Let $f(x) = x - 5\cos x = 0$; $f'(x) = 1 + 5\sin x$.

Let $x_{n+1} = x_n - \dfrac{x_n - 5\cos x_n}{1 + 5\sin x_n}$

n	x_n
1	-2
2	-1.97723545
3	-1.97081654
4	-1.97737132
5	-1.97735962
6	-1.97738302
7	-1.97738302

n	x_n
1	-4
2	-3.8470360
3	-3.8375083
4	-3.8374671
5	-3.8374671

The roots are approximately -3.83747 and -1.97738

21. Solve $x^3 - 7 = 0$. Let $f(x) = x^3 - 7$; $f'(x) = 3x^2$. Let $x_1 = 1.9$ and

$x_{n+1} = x_n - \dfrac{x_n^3 - 7}{3x_n^2}$

n	x_n
1	1.9
2	1.913019391
3	1.912931187
4	1.912931183

$\sqrt[3]{7} \approx 1.912931$

24. Solve $te^{-t/3} = 0.01$, or $e^{t/3} - 100t = 0$.

Let $g(t) = e^{t/3} - 100t = 0$; $g'(t) = \frac{1}{3}e^{t/3} - 100$. Let $t_1 = 20$ and

$t_{n+1} = t_n - \dfrac{e^{t_n/3} - 100t_n}{(1/3)e^{t_n/3} - 100}$

n	t_n
1	20
2	27.49875263
3	25.29170516
4	23.85234622
5	23.31736317
6	23.25581648
7	23.25508157
8	23.25508157

Give the second injection in about 23.26 hours (in about 2 hr, 15 min).

Problem Set 14.4   The Method of Successive Substitutions

3. Let $x_1 = 3$ and $x_{n+1} = 4 - \ell n x_n$.

   $x_{11} = 2.92627265$ and $x_{12} = 2.92627052$.   Root ≈ 2.9263.

6. (a) Let $x_1 = 0.85$ and $x_{n+1} = \cot x_n$

      $x_2 = 0.878477784$, $x_3 = 0.829241471$, $x_4 = 0.91594\ 777$,

      $x_5 = 0.767888109$, $x_6 = 1.035647948$, $x_7 = 0.592853809$.

   (b) Let $x_1 = 0.85$ and $x_{n+1} = \frac{1}{2}(x_n + \cot x_n)$

      $x_{11} = 0.860333083$ and $x_{12} = 0.860333776$.   Root ≈ 0.8603.

9. (a) $x^3 + 3x - 5 = 0$

      $x^3 + 3x = 5$

      $2(x^3 + 3x) = x^3 + 3x + 5$ [Added $x^3 + 3x$ to each side.]

      $2x(x^2 + 3) = x(x^2 + 3) + 5$

      $x = \frac{1}{2}[x + 5(x^2 + 3)^{-1}]$ [Divided each side by $2(x^2 + 3)$.]

   (b) Let $x_1 = 1$ and $x_{n+1} = \frac{1}{2}[x_n + 5(x_n^2 + 3)^{-1}]$.

      $x_9 = 1.15417121$ and $x_{10} = 1.15417144$.   Root ≈ 1.1542.

12. (a) Let $x = f(x) = x^{10} - 10x^4 + 9$.

$x^{10} = 10x^4 + x - 9$	$x^{10} - 10x^4 = x - 9$
$x = 10x^{-5} + x^{-8} - 9x^{-9}$	$x^4(x^6 - 10) = x - 9$
$x = g(x)$	$x = [(x-9)/(x^6-10)]^{1/4}$ near $x = 1$
	$x = h(x)$ near $x = 1$

   (b) $f'(x) = 10x^9 - 40x^3$, so $f'(1) = -30$.

      $g'(x) = -50x^{-6} - 8x^{-9} + 81x^{-10}$, so $g'(1) = 23$.

      $h'(x) = \frac{1}{4}[(x-9)(x^6-10)^{-1}]^{-3/4}(x^6-10)^{-2}(-5x^6+54x^5-10)$, so $h'(1) \approx 0.13$.

12. (c) $|h'(1)| < 1$, so let $x_1 = 1$ and $x_{n+1} = [(x_n-9)/(x_n^6-10)]^{1/4}$.

   $x_6 = 0.967135365$ and $x_7 = 0.967134962$. Root ≈ 0.9671.

15. $2cosx - 5x = 0$; $x = 0.4cosx$.

   Let $x_1 = 0.4$ and $x_{n+1} = 0.4cosx_n$

   $x_6 = 0.372557646$ and $x_7 = 0.372559765$.

   Root ≈ 0.3726.

18. Let $2te^{-0.4t} = 0.05$; $e^{0.4t} = 40t$; $0.4t = \ell n(40t)$; $t = 2.5\ell n(40t)$.

   Note that $C(16) \approx 0.053$ and $C(17) \approx 0.038$.

   Then let $t_1 = 16$, and $t_{n+1} = 2.5\ell n(40t_n)$

   $t_9 = 16.18193743$ and $x_{10} = 16.18193775$.

   Give the next injection in about 16.182 hours (about 16 hr, 10 min).

Problem Set 14.5   Numerical Solution of Differential Equations

 3. $\frac{dy}{dx} = ytanx$, so $y^{-1}dy = tanxdx$.

   Therefore, $\ell ny = \ell n(secx) + C$.

   $y(0) = 2 \Rightarrow C = \ell n2$, so $\ell ny = \ell n(secx) + \ell n2$; $y = 2secx = 2/cosx$.

   Now, let $h = 0.1$, $y_0 = y(0) = 2$, and $y_{k+1} = y_k + (y_k tanx_k)(0.1)$.

k	x_k	y_k	$y(x_k)$
0	0.0	2	2
1	0.1	2	2.0100
2	0.2	2.0201	2.0407
3	0.3	2.0610	2.0935
4	0.4	2.1248	2.1714
5	0.5	2.2146	2.2790
6	0.6	2.3356	2.4233
7	0.7	2.4954	2.6149
8	0.8	2.7056	2.8706
9	0.9	2.9841	3.2175
10	1.0	3.3602	3.7016

6. $y' + x^{-1}y = x^2$ [I.F. $= e^{\int 1/x \ dx} = x$]

$\frac{d}{dx}(xy) = x^3$, so $xy = \frac{1}{4}x^4 + C$

$y(2) = 1 \Rightarrow C = -2$, so $xy = \frac{1}{4}x^4 - 2$; $y = \frac{1}{4}x^3 - 2x^{-1}$.

Now, let $h = 0.2$, $y_0 = y(2) = 1$, $y_{k+1} = y_k + (x_k^2 - y_k/x_k)(0.2)$,

      and $\hat{y}_{k+1} = \hat{y}_k + 0.1[(x_k^2 - \hat{y}_k/x_k) + (x_{k+1}^2 - y_{k+1}/x_{k+1})]$.

k	x_k	y_k	$\hat{y}_k$	$y(x_k)$
0	2.0	1	1	1
1	2.2	1.7	1.7568	1.7529
2	2.4	2.5135	2.6321	2.6227
3	2.6	3.456	3.6415	3.6248
4	2.8	4.5422	4.7992	4.7737
5	3.0	5.7857	6.1190	6.0833
6	3.2	7.2	7.6140	7.5670
7	3.4	8.798	9.2973	9.2378
8	3.6	10.5925	11.1817	11.1084
9	3.8	12.596	13.2796	13.1917
10	4.0	14.8211	15.6036	15.5000

9. $y' + y = e^{-x}$ [I.F. $= e^{\int 1 dx} = e^x$]

$\frac{d}{dx}(e^x y) = 1$, so $e^x y = x + C$.

$y(0) = 01 \Rightarrow C = 0$, so $e^x y = x$; $y = xe^{-x}$.

Now, let $h = 0.2$, $y_0 = y(0) = 0$, $y_{k+1} = y_k + (-y_k + e^{-x_k})(0.2)$,

k	x_k	y_k	$y(x_k)$
0	0.0	0	0
1	0.2	0.2000	0.1637
2	0.4	0.3237	0.2681
3	0.6	0.3931	0.3293
4	0.8	0.4242	0.3595
5	1.0	0.4292	0.3679
6	1.2	0.4170	0.3614
7	1.4	0.3938	0.3452
8	1.6	0.3644	0.3230
9	1.8	0.3319	0.2975
10	2.0	0.2986	0.2707

3. $f(x) = sin^2x$, $h = \frac{3-1}{8} = 0.25$, partition is $\{1, 1.25, 1.5, 1.75, 2, 2.25,$
$\qquad\qquad\qquad\qquad\qquad\qquad\qquad\qquad\qquad\qquad 2.5, 2.75, 3\}$.

$\frac{0.25}{2}[sin^2(1) + 2sin^2(1.25) + 2sin^2(1.5) + 2sin^2(1.75) + 2sin^2(2) + 2sin^2(2.25)$
$\qquad\qquad + 2sin^2(2.5)] + 2sin^2(2.75) + sin^2(3)] \approx 1.29096107$

$f'(x) = 2sinxcosx = sin(2x)$; $f''(x) = 2cos(2x)$; $|f''(x)| \le 2$.

Therefore, $|E| \le \frac{(3-1)^3}{12(8)^2}(2) < 0.021$

6. $\left[-e^{1/x}\right]_1^3 = -e^{1/3} + e \approx 1.32266940$

$f(x) = x^{-2}e^{1/x}$, $h = \frac{3-1}{4} = 0.5$, partition is $\{1, 1.5, 2, 2.5, 3\}$.

$\frac{0.5}{3}[f(1) + 4f(1.5) + 2f(2) + 4f(2.5) + f(3)] \approx 1.35251944$

9. (a) $f(x) = cosx$, $f'(x) = -sinx$, $f''(x) = -cosx$, $f^{(3)}(x) = sinx$, so

$\qquad f(\pi/3) = 1/2$, $f'(\pi/3) = -\sqrt{3}/2$, $f''(\pi/3) = -1/2$, $f^{(3)}(\pi/3) = \sqrt{3}/2$.

$\qquad \therefore P_3(x) = (1/2) - (\sqrt{3}/2)(x-\pi/3) - (1/4)(x-\pi/3)^2 + (\sqrt{3}/12)(x-\pi/3)^3$.

(b) $cos(\pi/3 + 0.02) \approx (1/2) - (\sqrt{3}/2)(0.02) - (1/4)(0.02)^2 + (\sqrt{3}/12)(0.02)^3$
$\qquad\qquad\qquad\qquad\qquad\qquad\qquad\qquad \approx 0.48258065$

(c) $f^{(4)}(x) = cosx$, so $|f^{(4)}(x)| \le 1$.

$\qquad$ Therefore, $|E(x)| \le \frac{(0.02)^4}{4!}(1) < 0.000000007$.

12. Let $f(x) = 2x-1-cos2x$; $f'(x) = 2+2sin2x$.
$\qquad$ Let $x_1 = 0.6$ and

$x_{n+1} = x_n - \dfrac{2x_n - 1 - cos2x_n}{2 + 2sin2x_n}$

n	x_n
1	0.6
2	0.64201720
3	0.64171438
4	0.64171437

$\qquad$ Root ≈ 0.6417.

15. Let $e^{-x} = x-2$; $x = e^{-x}+2$.

Let $x_1 = 2.1$ and $x_{n+1} = e^{-x_n} + 2$

$x_5 = 2.120024045$ and $x_6 = 2.120028742$.

Root ≈ 2.1200.

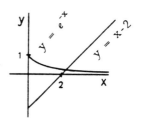

18. $x^3 + cosx = 6$; $x = \sqrt[3]{6-cosx}$.

Note that $1^3 + cos1 \approx 1.54$ and $2^3 + cos2 \approx 7.58$; a root is between 1 and 2.

Let $x_1 = 1.8$ and $x_{n+1} = \sqrt[3]{6-cosx_n}$

$x_5 = 1.84395297$ and $x_6 = 1.84395616$. Root ≈ 1.8440.

21. $f(x) = x - \frac{1}{2}x^2 + cosx$. Use Newton's Method on

$f'(x) = 1 - x - sinx = 0$; $f''(x) = -1 - cosx$.
Let $x_1 = 0.5$ and

$$x_{n+1} = x_n - \frac{1 - x_n - sinx_n}{-1 - cosx_n}$$

n	x_n
1	0.5
2	0.510957952
3	0.510973429
4	0.510973429

$f(x)$ is maximum at about $x = 0.5110$

24. $f(x) = (1+x^2)^{-1/2}$, $f'(x) = -x(1+x^2)^{-3/2}$, $f''(x) = (2x^2-1)(1+x^2)^{-5/2}$,

$f^{(3)}(x) = (9x-6x^3)(1+x^2)^{-7/2}$, $f^{(4)}(x) = (24x^4-72x^2+9)(1+x^2)^{-9/2}$, so

$f(0) = 1$; $f'(0) = 0$, $f''(0) = -1$, $f^{(3)}(0) = 0$; $f^{(4)}(0) = 9$.

Then $P_4(x) = 1 - \frac{1}{2}x^2 + \frac{3}{8}x^4$.

$(1.0009)^{-1/2} = [1+(0.03)^2]^{-1/2} = f(0.03) \approx P_4(0.03)$

$$= 1 - \frac{1}{2}(0.03) + \frac{3}{8}(0.03)^2 \approx 0.999550303$$

CHAPTER 15 INFINITE SERIES

Problem Set 15.1 Infinite Sequences and Series

3. (a) Converges to 0.   (b) Converges to -4.

   (c) $c_n = n^{2/3}$. It diverges.   (d) Converges to $\ln 2$.

6. (a) $\lim_{n \to \infty} \dfrac{4 - 1/n}{1 + 60/n} = \dfrac{4-0}{1+0} = 4$

   (b) $\lim_{n \to \infty} \dfrac{2/n^3 - 1}{2 + 1/n^2 - 4/n^3} = \dfrac{0-1}{2+0-0} = -\dfrac{1}{2}$

   (c) $\lim_{n \to \infty} \dfrac{3}{2/n\sqrt{n} + 1} = \dfrac{3}{0+1} = 3$

   (d) $\lim_{n \to \infty} \dfrac{10/n}{1 + 1/n^2} = \dfrac{0}{1+0} = 0$

   (e) $\lim_{n \to \infty} \left[\dfrac{1 + 3/n^2}{1/n^3 - 8}\right]^{1/3} = \left[\dfrac{1+0}{1-8}\right]^{1/3} = -\dfrac{1}{2}$

   (f) $\lim_{n \to \infty} \dfrac{\sqrt[3]{1 + 3/n^2}}{1/n + 2} = \dfrac{\sqrt[3]{1+0}}{0+2} = \dfrac{1}{2}$ [Divided numerator by $\sqrt[3]{n^3}$ and denominator by n.]

9. $[\frac{1}{4} - \frac{1}{9}] + [\frac{1}{9} - \frac{1}{16}] + [\frac{1}{16} - \frac{1}{25}] + \cdots$

   $S_2 = \frac{1}{4} - \frac{1}{9}$, $S_3 = \frac{1}{4} - \frac{1}{16}$, $S_4 = \frac{1}{4} - \frac{1}{25}$, $\ldots$, $S_n = \frac{1}{4} - \dfrac{1}{(n+1)^2} \to \frac{1}{4}$

12. $\lim_{n \to \infty} a_n \neq 0$, so the series diverges.

15. $\lim_{n \to \infty} \ln(2 + \frac{1}{n}) = \ln 2 \neq 0$, so the series diverges.

18. $\lim_{n \to \infty} 2^{1/n} \neq 0$ (since each term is greater than 1); the series diverges.

Problem Set 15.1   159

21. The partial sums of this series are one-half of the corresponding partial sums of the harmonic series, so the partial sums go to infinity. Thus, the series diverges.

24. The partial sums of this series are three times the corresponding partial sums of the series of Example G, so these partial sums are less than 6. Since they are bounded, the series converges.

27. (a) $a_n = \dfrac{3}{1 + 2/n^2}$, so it converges to 3.

    (b) The sequence is 1, 3, 1, 3,..., so it diverges.

    (c) $\ln(2.7) < 1$, so $[\ln(2.7)]^n$ converges to 0.

    (d) This is the sequence of partial sums of the harmonic series, so it diverges.

    (e) It converges to 1.

    (f) $a_n = \dfrac{2^n - 1}{2^{n-1}} = \dfrac{2 - 1/2^{n-1}}{1}$, so it converges to 2.

30. $S_6 = 3 + 5 + 11 + 29 + 83 + 245 = 376$

33. (a) 2, 3, 5, 8, 13, 21, 34, 55, 89

    (b) The successive ratios are 1, 2, 1.5, 1.667, 1.6, 1.625, 1.6154, 1.6190, 1.6176, 1.6182, 1.6180, 1.6181, 1.6180. It converges to 1.618 (to three decimal places).

36. Let the amount of air originally in the cylinder be 1 unit. Let a_n be the amount of air left after n strokes, $n = 1, 2, 3, ...$

    Then $a_n = (1/3)^n$. $(1/3)^n < 1/1000$ if $n > (\ln 1000)/(\ln 3) \approx 6.3$, so it will take 7 strokes.

3. $\dfrac{3[1 - (-1/2)^{12}]}{1 - (-1/2)} = \dfrac{4095}{2048} \approx 1.99951172$

6. $\dfrac{100[1 - (1.025)^{12}]}{1 - 1.025} \approx 1379.25530$

9. It is a geometric series with n = 20, r = 1.04, a = 500.

$\dfrac{500[(1.04)^{20} - 1]}{1.04 - 1} = \$14{,}889.04$

12. It is a geometric series with n = 5, r = $(1.07)^{-1}$, a = $20000(1.07)^{-1}$.

$\dfrac{20000(1.07)^{-1}\{[(1.07)^{-1}]^{5} - 1\}}{(1.07)^{-1} - 1} = \$82{,}003.95$

15. Geometric series with a = 1, r = $\frac{2}{3}$, so its sum is $\dfrac{1}{1 - 2/3} = 3$.

18. Geometric series with a = 1, r = $\ell n 2$, so its sum is $\dfrac{1}{1 - \ell n 2} \approx 3.2589$.
    [Note that $\ell n 2 < 1$.]

21. Each term determines a converging geometric series. The sum is

$\dfrac{1}{1 - 0.54} + \dfrac{1}{1 - (-0.2)} \approx 3.00724638$

24. $0.71 + 0.0071 + 0.000071 + \cdots$ is a geometric series with a = 0.71
    and r = 0.01, so the sum is $\dfrac{0.71}{1 - 0.01} = \dfrac{71}{99}$.

27. Let the area of the triangle be 1 unit. Then the area of the largest
    square is 1/2. The area of each subsequent square is one-fourth of
    the area of its predecessor. The sum of the areas is a geometric
    series with a = 1/2 and r = 1/4, so the sum is $\dfrac{1/2}{1 - 1/4} = \dfrac{2}{3}$.

    That is, two-thirds of the area of the triangle is represented.

30. The present value of the perpetuity is a geometric series with $a = 50000(1.075)^{-1}$ and $r = (1.075)^{-1}$.

Its sum is $\dfrac{50000(1.075)^{-1}}{1 - (1.075)^{-1}} = \$666{,}666.67$.

33. (a) $\dfrac{36(0.01)}{1 - 0.01} = \dfrac{4}{11}$

(b) $\dfrac{0.36}{1 - 0.36} = \dfrac{9}{16}$

(c) $\dfrac{\sqrt{5}/3}{1 - \sqrt{5}/3} = \dfrac{\sqrt{5}}{3 - \sqrt{5}}$

(d) $\dfrac{3/4}{1 - (-3/4)} = \dfrac{3}{7}$

36. $40000 = \dfrac{a[1 - (1.005)^{180}]}{1 - 1.005}$, so $a = \dfrac{40000(0.005)}{(1.005)^{180} - 1} = \$137.55$

39. At beginning of 1st month: 900
    At beginning of 2nd month: $900(3/4) + 900$
    At beginning of 3rd month: $[900(3/4)](3/4) + 900(3/4) + 900$

A geometric series is formed with $a = 900$ and $r = 3/4$ (reverse terms).

So in the long run there will be $\dfrac{900}{1 - 3/4} = 3600$ fish at the beginning of the month (and 2700 at the end of the month).

Problem Set 15.3   Positive Series

3. $a_k \geq 1/k$, so the series diverges.   (Comparison Test)

6. $a_k = \dfrac{1 + 1/k}{k} > \dfrac{1}{k}$, so the series diverges.   (Comparison Test)

9. $\dfrac{2^{k+1}}{(k+1)^3}\,\dfrac{k^3}{2^k} = \dfrac{2k^3}{(k+1)^3} = \dfrac{2}{(1 + 1/k)^3} \to 2$.   Series diverges.

12. $\dfrac{2^{k+1} + 3^{k+1}}{4^{k+1}}\,\dfrac{4^k}{2^k + 3^k} = \dfrac{2/3^k + 3/2^k}{4(1/3^k + 1/2^k)} \to 0$.   Series converges.

15. $a_k = 1/k^{1/2}$. $p = 1/2 < 1$, so the series diverges.

18. $a_k = 1/k^{1.01}$. $p = 1.01 > 1$, so the series converges.

21. $\int_1^\infty 2(x+1)^{-1.2}dx = \lim_{b\to\infty} \left[-10(x+1)^{-0.2}\right]_1^\infty = 10(2)^{-0.2}$, so series converges.

24. $\frac{\sqrt{k+1}}{k^{3/2}} > \frac{\sqrt{k}}{k^{3/2}} = \frac{1}{k}$, so the series diverges. (Comparison Test)

27. $a_k = \frac{1 - 2/k^2}{1 + 1/k^2} \to 1$, so the series diverges. (Nth Term Test)

30. $a_k > 1/\sqrt{k} = 1/k^{1/2}$, so the series diverges. (Comparison Test using a
$\qquad\qquad\qquad\qquad\qquad\qquad\qquad$ p-series with $p = 1/2$ -- Example E)

33. It converges. (Geometric series with $r = -1/2$)

36. It converges since $\int_2^\infty x^{-1}(\ln x)^{-2}dx = (\ln 2)^{-1}$. (Integral Test)

39. It equals $4\sum(\frac{1}{3})^k - 2\sum(k^{-2})$. The first is a geometric series with $r = \frac{1}{3}$

and the second is a p-series with $p = 2$ (Example E). Each converges.

42. $a_k < 4/k^2$. $\sum(4/k^2)$ converges since $\sum(1/k^2)$ does (p-series with $p = 2$).
$\qquad\qquad\qquad\qquad$ Thus, the series converges by the Comparison Test.

45. The series converges. (Geometric series with $r = 1/(1.1) < 1$)

48. (Ratio Test) $\frac{2^{k+1}}{(k+1)!} \frac{k!}{2^k} = \frac{2}{k+1} \to 0 < 1$, so the series converges.

51. (Ratio Test) $\frac{3^{k+2}[2(k+1)+1]}{(k+1)4^{k+1}} \frac{4^k k}{3^{k+1}(2k+1)} = \frac{3(2 + 3/k)}{4(1 + 1/k)(2 + 1/k)} \to \frac{3}{4} < 1$,
so the series converges.

54. (Ratio Test) $\frac{(k+1)!}{3^{k+1}} \frac{3^k}{k!} = \frac{k+1}{3} \to \infty$, so the series diverges.

57. $(1 + 1/k)^k > 1$, so the series diverges. (Nth Term Test)

3. $\sin(k\pi/3)$ has only values of $-\sqrt{3}/2$, $\sqrt{3}/2$, and 0.   The absolute value of each is less than 1.   Therefore, $|a_k| < 1/k^2$, so the series converges absolutely by the Comparison Test, using the p-series with $p = 2$.   Hence, the original series converges.

6. The terms are alternating, decreasing in absolute value, and converging to 0, so the series converges.

9. The terms alternate and converge to 0.   Also, $\dfrac{a_{k+1}}{a_k} = \dfrac{1 + 1/k}{2} \leq 1$, so $a_{k+1} < a_k$.   Therefore, the series converges.

12. Converges absolutely since $\sum(4/k^{1.1})$ is 4 times the p-series with $p = 1.1$.

15. Converges absolutely by ratio test. $\dfrac{(k+1)^4}{2^{k+1}} \dfrac{2^k}{k^4} = \dfrac{(1 + 1/k)^4}{2} \to \dfrac{1}{2} < 1$

18. Diverges by Nth Term Test since $a_k \to \sin 1 \neq 0$.

21. $\left| \dfrac{2^{k+1}x^{k+1}}{(k+1)!} \dfrac{k!}{2^k x^k} \right| = \dfrac{2|x|}{k+1} \to 0$ for all x, so series converges on $(-\infty, \infty)$.

24. $\left| \dfrac{(k+1)^2 x^{k+1}}{k^2 x^k} \right| = \dfrac{(k+1)^2 |x|}{k^2} \to |x| < 1$ for x in $(-1,1)$.

   At end points: Both $\sum(-1)^k k^2$ and $\sum k^2$ diverge, so the interval of convergence is $(-1,1)$.

27. $\left| \dfrac{3^{k+1}x^{k+1}}{(k+1)!} \dfrac{k!}{3^k x^k} \right| = \dfrac{3|x|}{k+1} \to 0 < 1$ for all x in $(-\infty, \infty)$.

30. $e^2 = 1 + 2 + \dfrac{2^2}{2!} + \dfrac{2^3}{3!} + \dfrac{2^4}{4!} + \cdots$

33. $\dfrac{1}{1-3x} = 1 + (3x) + (3x)^2 + (3x)^3 + (3x)^4 + \cdots$ for $-1 < 3x < 1$, for x in $(-1/3, 1/3)$.

36. $e^{-x} = 1 + (-x) + \dfrac{(-x)^2}{2!} + \dfrac{(-x)^3}{3!} + \cdots$

$\qquad = 1 - x + \dfrac{x^2}{2!} - \dfrac{x^3}{3!} + \cdots,$ for x any real number.

39. $e^{2x} = 1 + (2x) + \dfrac{(2x)^2}{2!} + \dfrac{(2x)^3}{3!} + \cdots$

$\qquad = 1 + 2x + \dfrac{4x^2}{2!} + \dfrac{8x^3}{3!} + \cdots,$ for x any real number.

42. $\dfrac{cosx}{1-x} = (1 - \dfrac{x^2}{2!} + \dfrac{x^4}{4!} - \cdots)(1 + x + x^2 + x^3 + x^4 + \cdots)$

$\qquad = 1 + x + \tfrac{1}{2}x^2 + \tfrac{1}{2}x^3 + \tfrac{13}{24}x^4 + \cdots,$ for x in (-1,1).

45. $sin2x + cos(x^2) = (2x - \dfrac{(2x)^3}{3!} + \cdots) + (1 - \dfrac{x^4}{2!} + \cdots)$

$\qquad\qquad = 1 + 2x - \tfrac{4}{3}x^3 - \tfrac{1}{2}x^4 + \cdots,$ for x any real number.

48. $\displaystyle\int_0^{0.5} sin(x^2)dx = \int_0^{0.5} [x^2 - \tfrac{1}{6}x^6 + \tfrac{1}{120}x^{10} - \tfrac{1}{5040}x^{14} + \cdots]dx$

$\left[\tfrac{1}{3}x^3 - \tfrac{1}{42}x^7 + \tfrac{1}{1320}x^{11} - \tfrac{1}{75600}x^{15} + \cdots \right]_0^{0.5} \approx 0.041481$

51. First solve $3 = \dfrac{1+x}{1-x}$ for x; obtain $x = \tfrac{1}{2}$.

$\therefore \ \ell n3 = \ell n \dfrac{1 + 1/2}{1 - 1/2} = 2(\tfrac{1}{2}) + \tfrac{2}{3}(\tfrac{1}{2})^3 + \tfrac{2}{5}(\tfrac{1}{2})^5 + \tfrac{2}{7}(\tfrac{1}{2})^7 + \tfrac{2}{9}(\tfrac{1}{2})^9 + \tfrac{2}{11}(\tfrac{1}{2})^{11} + \cdots$

$\qquad\quad \approx 1.0986$

54. $x^2e^{-x} + (1-x)^{-1}$

$\quad = x^2(1 - x + \dfrac{x^2}{2!} - \dfrac{x^3}{3!} + \dfrac{x^4}{4!} - \dfrac{x^5}{5!} + \cdots) + (1+x+x^2+x^3+x^4+x^5+ \cdots)$

$\quad = 1 + x + 2x^2 + \tfrac{3}{2}x^4 + \tfrac{5}{6}x^5 + \tfrac{25}{24}x^6 + \cdots$

Chapter 15 Review Problem Set

3. 3, 10, 24, 52, 108

6. 0.027 + 0.000027 + 000000027 + $\cdots$ is a geometric series with
   a = 0.027 and r = 0.001, so the sum is $\dfrac{0.027}{1 - 0.001} = \dfrac{27}{999}$.

9. The series converges. (Geometric series with r = $e-2 \approx 0.72$)

12. Diverges by the Nth Term Test since $a_k \to cos3 \neq 0$.

15. Diverges by Comparison Test since $a_k \geq \dfrac{k}{k^2 + k^2} = \dfrac{1}{2k}$.
    (one-half times Harmonic Series terms)

18. Converges by Alternating Series Test. a_k alternates and its absolute
    value is decreasing and converging to 0.

21. Converges by Comparison Test since $a_k < 1/2^k$. (Geometric Series)

24. Converges by Comparison Test since $a_k < 1/e^k$. (Geometric Series)

27. (a) See Problem 39 of Section 4.

    (b) $(x+2)cosx = (2+x)(1 - \dfrac{x^2}{2!} + \cdots) = 2 + x - x^2 - \frac{1}{2}x^3 + \cdots$, for x
    any real number.

30. Each term determines a converging geometric series. The sum is

    $\dfrac{3}{1 - 5/6} - \dfrac{21/2}{1 - (-3/4)} = 12$.

33. $10 + 10(0.60) + 10(0.60)^2 + \cdots$ is a geometric series with a = 10 and
    r = 0.60. Its sum is $\dfrac{10}{1 - 0.60} = 25$ mg.

36. (a) $(x+1)\ell n(x+1) = (1+x)(x - \frac{1}{2}x^2 + \frac{1}{3}x^3 - \frac{1}{4}x^4 + \cdots)$

$$= x + \frac{1}{2}x^2 - \frac{1}{6}x^3 + \frac{1}{12}x^4 + \cdots$$

Then $\int_0^{0.5} (x+1)\ell n(x+1)\ dx \approx \int_0^{0.5}(x + \frac{1}{2}x^2 - \frac{1}{6}x^3 + \frac{1}{12}x^4)\,dx$

$$= \left[\frac{1}{2}x^2 + \frac{1}{6}x^3 - \frac{1}{24}x^4 + \frac{1}{60}x^5\ \right]_0^{0.5} = 0.14375$$

(b) Let $f(x) = (x+1)\ell n(x+1)$, $h = 0.1$, partition be
$$\{0,0.1,0.2,0.3,0.4,0.5\}.$$

$$\frac{0.1}{2}[f(0) + 2f(0.1) + 2f(0.2) + 2f(0.3) + 2f(0.4) + f(0.5)]$$
$$\approx 0.143986057$$

(c) $\int_0^{0.5} (x+1)\ell n(x+1)\ dx = \left[\frac{1}{2}(x+1)^2\ell n(x+1) - \frac{1}{4}(x+1)^2\right]_0^{0.5}$ [Formula 66]

$$\approx 0.143648246$$

Problem Set 16.1   Discrete Distributions

3. S = {HHHH,HHHT,HHTH,HHTT,HTHH,HTHT,HTTH,HTTT,THHH,THHT,THTH,THTT,TTHH,
TTHT,TTTH,TTTT}

$\text{prob}\{X = 3\} = \text{prob}\{HHHT,HHTH,HTHH,THHH\} = 4/16$
$\text{prob}\{X \le 1\} = \text{prob}\{TTTT,HTTT,THTT,TTHT,TTTH\} = 5/16$

6.
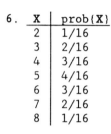

X	prob{X}
2	1/16
3	2/16
4	3/16
5	4/16
6	3/16
7	2/16
8	1/16

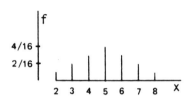

9. $\sigma^2 = [0(1/16) + 1(4/16) + 4(6/16) + 9(4/16) + 16(1/16)] - (2)^2 = 1$

12. (Expected winnings for red) + (Expected winnings for blue)
                              + (Expected winnings for white)
   = ($3)(6/20) + ($10)(2/20) + (-$2)(12/20) = $0.70

15. $\text{prob}\{X = 3\} = \text{prob}\{(1,1,1)\} = 1/216$
$\text{prob}\{X = 4\} = \text{prob}\{(1,1,2),(1,2,1),(2,1,1)\} = 3/216$
$\text{prob}\{X > 4\} = 1 - \text{prob}\{X \le 4\} = 1 - \text{prob}\{X = 3,4\} = 1 - 4/216 = 212/216$

18. (a) $\dfrac{e^{-p}p^0}{0!} + \dfrac{e^{-p}p^1}{1!} + \dfrac{e^{-p}p^2}{2!} + \cdots = e^{-p}(1 + p + p^2/2! + p^3 + \cdots)$

   $= e^{-p}e^{p} = 1$

   (b) $\mu = \dfrac{0e^{-p}p^0}{0!} + \dfrac{1e^{-p}p^1}{1!} + \dfrac{2e^{-p}p^2}{2!} + \dfrac{3e^{-p}p^3}{3!} + \cdots$

   $= e^{-p}p(0 + 1 + p + p^2/2! + p^3/3! + \cdots) = e^{-p}pe^{p} = p$

21.

				X	prob{X}
HHHHH	HTHHH	THHHH	TTHHH	0	1/32
HHHHT	HTHHT	THHHT	TTHHT	1	5/32
HHHTH	HTHTH	THHTH	TTHTH	2	10/32
HHHTT	HTHTT	THHTT	TTHTT	3	10/32
HHTHH	HTTHH	THTHH	TTTHH	4	5/32
HHTHT	HTTHT	THTHT	TTTHT	5	1/32
HHTTH	HTTTH	THTTH	TTTTH		
HHTTT	HTTTT	THTTT	TTTTT		

$$\mu = 0(1/32) + 1(5/32) + 2(10/32) + 3(10/32) + 4(5/32) + 5(1/32) = 2.5$$

$$\sigma^2 = [0(1/32) + 1(5/32) + 4(10/32) + 9(10/32) + 16(5/32) + 25(1/32)]$$
$$- (2.5)^2 = 1.25$$

Problem Set 16.2 Continuous Distributions

3. $k(x^2-2x) = kx(x-2) > 0$ on $[3,4]$ if $k > 0$.

$$1 = \int_3^4 k(x^2-2x)dx = k\left[\frac{1}{3}x^3-x^2\right]_3^4 = 20k, \text{ so } k = 0.05.$$

6. $k\sqrt{x} \geq 0$ on $[0,4]$ if $k > 0$.

$$1 = \int_0^4 kx^{1/2}dx = k\left[\frac{2}{3}x^{3/2}\right]_0^4 = 16k/3, \text{ so } k = 3/16$$

9. 45 minutes = 0.75 hours. $\text{prob}\{X \geq 0.75\} = \int_{0.75}^1 \frac{4}{3}(1-x^3)dx = 27/256.$

12. (a) $f(x) = \frac{1}{3}$ on $[1,4]$ and $f(x) = 0$ elsewhere.

(b) $\mu = \int_1^4 x(\frac{1}{3})dx = 2.5$

$$\sigma^2 = \int_1^4 x^2(\frac{1}{3})dx - (2.5)^2 = 7 - 6.25 = 0.75.$$

15. $\mu = \int_0^1 x[\frac{4}{3}(1-x^3)]dx = 0.4$ hours (24 minutes)

18. $\text{prob}\{0.6 < Y < 0.7\} = \text{prob}\{0.6 < \ell n X < 0.7\} = \text{prob}\{e^{0.6} < X < e^{0.7}\}$

$= \frac{1}{6}(e^{0.7} - e^{0.6}) \approx 0.03194$

21. Equations of the line segments are $y = \frac{1}{6}(t-5)$ on $[5,8]$; $y = -\frac{1}{2}(t-9)$ on $[8,9]$.

$\mu = \int_5^8 t[\frac{1}{6}(t-5)]dt + \int_8^9 t[-\frac{1}{2}(t-9)]dt \approx 7.3333$ (at 7:20 p.m.)

Problem Set 16.3 The Exponential and Normal Distributions

3. $\mu = \frac{1}{2}$ means $k = 2$, so $f(t) = 2e^{-2t}$ (1st theorem of section)

$\text{prob}\{T > 1\} = \int_1^\infty 2e^{-2t}dt = e^{-2} \approx 0.13534$

6. $\text{prob}\{T \geq 10\} = \int_{10}^\infty \frac{1}{6}e^{-t/6}dt = e^{-5/3} \approx 0.18888$

9. $\text{prob}\{-0.5 \leq Z \leq 1.5\}$
   $= \text{prob}\{-0.5 \leq Z \leq 0\} + \text{prob}\{0 \leq Z \leq 1.5\}$
   $= \text{prob}\{0 \leq Z \leq 0.5\} + 0.4332$
   $= 0.1915 + 0.4332$
   $= 0.6247$

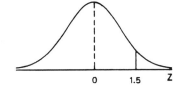

$\text{prob}\{Z > 1.5\}$
$= \text{prob}\{Z \geq 0\} - \text{prob}\{0 \leq Z \leq 1.5\}$
$= 0.5000 - 0.4332$
$= 0.0668$

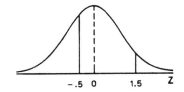

12. $z_{0.05} = \dfrac{0.05 - 0.08}{0.02} = -1.5$ and $z_{0.12} = \dfrac{0.12 - 0.08}{0.02} = 2$

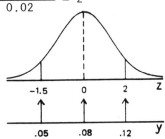

prob$\{0.05 \leq Y \leq 0.12\}$
= prob$\{-1.5 \leq Z \leq 2\}$
= prob$\{-1.5 \leq Z \leq 0\}$ + prob$\{0 \leq Z \leq 2\}$
= prob$\{0 \leq Z \leq 1.5\}$ + 0.4772
= 0.4332 + 0.4772 = 0.9104

$z_{0.07} = \dfrac{0.07 - 0.08}{0.02} = -0.5$

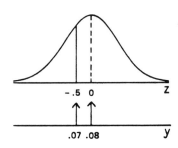

prob$\{Y \leq 0.07\}$
= prob$\{Z \leq -0.5\}$
= prob$\{Z \leq 0\}$ - prob$\{-0.5 \leq Z \leq 0\}$
= 0.5 - prob$\{0 \leq Z \leq 0.5\}$
= 0.5 - 0.1915
= 0.3085

15. Let **X** denote the number of ounces in 12-ounce bottles of the soda.

$z_{12} = \dfrac{12 - 12.1}{0.2} = -0.5$

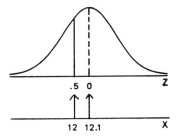

prob$\{X \geq 12\}$
= prob$\{Z \geq -0.5\}$
= prob$\{-0.5 \leq Z \leq 0\}$ + prob$\{Z \geq 0\}$
= prob$\{0 \leq Z \leq 0.5\}$ + 0.5
= 0.1915 + 0.5000
= 0.6915, so about 69% of the bottles.

18. (a) **X** has a standard normal distribution.

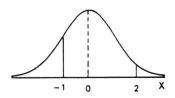

prob$\{-1 \leq X \leq 2\}$
= prob$\{-1 \leq X \leq 0\}$ + prob$\{0 \leq X \leq 2\}$
= prob$\{0 \leq X \leq 1\}$ + 0.4772
= 0.3413 + 0.4772
= 0.8185

(b) It is the mean of the standard normal variable. It equals 0.
(c) It is the variance of the standard normal variable plus the square of the mean, so it equals $1^2 + 0^2 = 1$.
(d) It involves a normal distribution with $\mu = 1$ and $\sigma = 2$.

Let $z = \dfrac{x-1}{2}$, so $dz = \tfrac{1}{2}dx$.

Then the integral is $(1/\sqrt{2\pi})\displaystyle\int_{-\infty}^{\infty} 4z^2 e^{-z^2/2}dz = 4(1)^2 = 4$ [from (c)]

21. Let **X** denote the number of pieces of
mail handled by the post office in
a day.

$$z_{13200} = \frac{13200 - 12500}{500} = 1.4$$

prob$\{X > 13200\}$
$= $ prob$\{Z > 1.4\}$
$= $ prob$\{Z \geq 0\} - $ prob$\{0 \leq Z \leq 1.4\}$
$= 0.5 - 0.4192$
$= 0.0808$

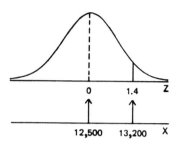

Problem Set 16.4   Continuous Bivariate Distributions

3. (a) prob$\{Y \geq 1\} = \dfrac{(1/2)(2)(1)}{4}$

$= 0.25$

(b) prob$\{Y \geq X/4\} = \dfrac{4 - (1/2)(4)(1)}{4}$

$= 0.5$

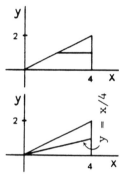

6. (a) $1 = \displaystyle\int_0^4 \int_0^y kxy \ dydx$

$= 32k$, so $k = 1/32$.

(b) prob$\{Y \geq 2\} = \displaystyle\int_2^4 \int_0^y (1/32)xy \ dydx$

$= 15/16$

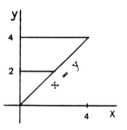

9. $\mu_x = 4$, so $k_x = 1/4$; $\mu_y = 8$, so $k_y = 1/8$.

Therefore, $f(x,y) = (\frac{1}{4}e^{-x/4})(\frac{1}{8}e^{-y/8})$

prob$\{X+Y \leq 10\} = \displaystyle\int_0^{10} \int_0^{10-x} (\frac{1}{4}e^{-x/4})(\frac{1}{8}e^{-y/8}) \, dydx$

$= 1 + e^{-5/2} - 2e^{-5/4} \approx 0.50908$

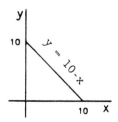

12. $\text{prob}\{X > Y\} = \int_0^\infty \int_0^x (\frac{1}{10}e^{-x/10})(\frac{1}{6}e^{-y/6})\,dy\,dx$

   $= 5/8$

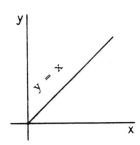

15. (a) $\text{Area}(D) = 2\int_0^4 (16-x^2)\,dx = 256/3,$

   so $f(x,y) = 3/256$ on D
   and $f(x,y) = 0$ elsewhere.

   (b) $\text{prob}\{X+Y \geq 4\} = \int_{-3}^4 \int_{4-x}^{16-x^2} (3/256)\,dy\,dx$

   $= 343/512 \approx 0.66992$

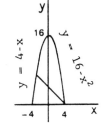

18. Uniform density function: $f(x,y) = 1$

   $\text{prob}\{Y \leq 4X\} = \dfrac{1 - (1/2)(1/4)(1)}{1} = \dfrac{7}{8}$

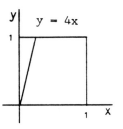

Chapter 16 Review Problem Set

3. $\mu = 1(1/2) + 2(1/4) + 3(1/6) + 4(1/12) = 11/6$
   $\sigma^2 = [1(1/2) + 4(1/4) + 9(1/6) + 16(1/12)] - (11/6)^2 = 35/36$

6.

X	Y	prob{X}
1	1	1/6
2	0	1/6
3	1	1/6
4	4	1/6
5	9	1/6
6	16	1/6

$\mu_y = 0(1/6) + 1(2/6) + 4(1/6) + 9(1/6) + 16(1/6)$
$= 31/6$

$\sigma_y^2 = [0(1/6)+1(2/6)+16(1/6)+81(1/6)+256(1/6)]$
$\qquad\qquad - (31/6)^2 = 1169/36 \approx 32.4722$

9. $\text{prob}\{T > 5\} = \int_5^{10} 62.5t(25+t^2)^{-2}dt = 0.375$

12. $1 = \int_0^\infty ke^x(1+e^x)^{-2}dx = k/2$, so $k = 2$.

$\text{prob}\{X \le 2\} = \int_0^2 2e^x(1+e^x)^{-2}dx = 1 - 2(1+e^2)^{-1} \approx 0.76159$

15. $z_{0.6} = \dfrac{0.6 - 0}{2} = 0.3$ and $z_{1.4} = \dfrac{1.4 - 0}{2} = 0.7$

$\text{prob}\{0.6 \le X \le 1.4\}$
$= \text{prob}\{0.3 \le Z \le 0.7\}$
$= \text{prob}\{0 \le Z \le 0.7\} - \text{prob}\{0 \le Z \le 0.3\}$
$= 0.2580 - 0.1179$
$= 0.1401$

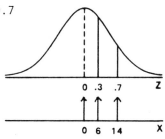

$z_{-0.6} = \dfrac{-0.6 - 0}{2} = -0.3$ and $z_{2.6} = \dfrac{2.6 - 0}{2} = 1.3$

$\text{prob}\{-0.6 \le X \le 2.6\}$
$= \text{prob}\{-0.3 \le Z \le 1.3\}$
$= \text{prob}\{-0.3 \le Z \le 0\} + \text{prob}\{0 \le Z \le 1.3\}$
$= \text{prob}\{0 \le Z \le 0.3\} + 0.4032$
$= 0.1179 + 0.4032$
$= 0.5211$

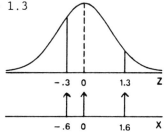

18. $z_{58} = \dfrac{58 - 62}{4.5} = -0.89$

$\text{prob}\{X \le 52\}$
$= \text{prob}\{Z \le -0.89\}$
$= \text{prob}\{Z \le 0\} - \text{prob}\{-0.89 \le Z \le 0\}$
$= 0.5 - \text{prob}\{0 \le Z \le 0.89\}$
$= 0.5 - 0.3133$
$= 0.1867$

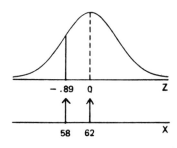

21. Pitcher: $\mu_p = 8$
    Catcher: $\mu_c = 6$

$f(p,c) = (\frac{1}{8}e^{-p/8})(\frac{1}{6}e^{-c/6})$

prob{at least one late}
= 1 - prob{both are on time}

$= 1 - \int_0^{10}\int_0^{10} (\frac{1}{8}e^{-p/8})(\frac{1}{6}e^{-c/6})\,dp\,dc$

$= 1 - (1-e^{-1.25})(1-e^{-5/3}) \approx 0.42127$

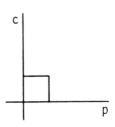

24. (a) Area(D) $= \int_0^1 [(4-x^2)-(3x)]\,dx = 13/6$, so

    $f(x,y) = 6/13$ on D; $f(x,y) = 0$ elsewhere.

   (b) $\mu_x = \int_0^1\int_{3x}^{4-x^2} x(6/13)\,dy\,dx = 9/26 \approx 0.346$

   $\mu_y = \int_0^1\int_{3x}^{4-x^2} y(6/13)\,dy\,dx = 158/65 \approx 2.431$

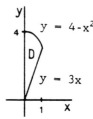

27. (a) $1 = \int_0^1\int_0^x kye^{x^3}\,dy\,dx = k(e-1)/6$, so

    $k = 6/(e-1)$

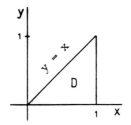

   (b) prob$\{Y \le \frac{1}{2}X\} = \int_0^1\int_0^{x/2} [6/(e-1)]e^{x^3}\,dy\,dx$

    $= 0.25$

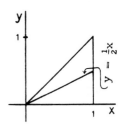